Harm reduction in nicotine addiction

Helping people who can't quit

A report by the Tobacco Advisory Group of the Royal College of Physicians, October 2007

Royal College of Physicians
Setting higher medical standards

Acknowledgements

The members of the Tobacco Advisory Group acknowledge with thanks Diana Beaven, Hannah Thompson and Joanna Reid of the RCP Publications Department for the editing and production of the book.

Citation of this report: Royal College of Physicians. *Harm reduction in nicotine addiction: helping people who can't quit.* A report by the Tobacco Advisory Group of the Royal College of Physicians. London: RCP, 2007.

Royal College of Physicians of London
11 St Andrew's Place, London NW1 4LE
www.rcplondon.ac.uk
Registered Charity No 210508

ISBN 978-1-86016-319-7

Cover photograph: Brenda Ann Kenneally
Design: Suzanne Fuzzey

Typeset by Dan-Set Graphics, Telford, Shropshire
Printed in Great Britain by The Lavenham Press Ltd, Suffolk

Contents

Contributors

Deborah Arnott *Director, Action on Smoking and Health, London*

Richard Ashcroft *Professor of Bioethics, School of Law, Queen Mary, University of London*

David Balfour *Professor of Behavioural Pharmacology, Division of Pathology and Neuroscience, University of Dundee*

Neal Benowitz *Professor of Medicine, Psychiatry and Biopharmaceutical Sciences, University of California, San Francisco*

John Britton *Professor of Epidemiology, University of Nottingham*

Paul Clarke *Professor of Pharmacology, McGill University, Montreal*

Richard Edwards *Senior Lecturer in Epidemiology, Department of Public Health, University of Otago, Wellington, New Zealand*

Jonathan Foulds *Professor of Health Education and Behavioral Science and Director, Tobacco Dependence Program, UMDNJ School of Public Health, New Brunswick*

Anna Gilmore *Reader in Public Health, University of Bath*

Allan Hackshaw *Deputy Director, Cancer Research UK and UCL Cancer Trials Centre, University College London*

Jack Henningfield *Professor of Behavioral Biology, Johns Hopkins University School of Medicine; Consultant, Pinney Associates*

Richard Hubbard *Professor of Respiratory Epidemiology, University of Nottingham*

Lynn Kozlowski *Professor of Health Behavior, University at Buffalo, State University of New York*

Ann McNeill *Professor of Health Policy and Promotion, University of Nottingham*

Members of the Tobacco Advisory Group of the Royal College of Physicians

John Britton (Chair)

Deborah Arnott

Tim Coleman

Linda Cuthbertson

Richard Edwards

Christine Godfrey

Allan Hackshaw

Martin Jarvis

Ann McNeill

Jennifer Percival

Mike Ward

Foreword

I am proud that the Royal College of Physicians remains at the forefront of policy development in the field of smoking 45 years after our influential intervention of 1962. John Britton and his team have produced another stimulating and radical report that faces up to the issue of nicotine addiction and challenges the current position on alternative nicotine products. I congratulate them on this comprehensive, carefully argued report and commend it to you.

October 2007   **Ian Gilmore**
President, Royal College of Physicians

Preface

Cigarette smoking is powerfully addictive, and caused 100 million deaths in the 20th century. In the 21st century, if smoking trends persist as expected, one billion people will die from smoking tobacco. All of these deaths are preventable.

Current national and international tobacco control policies focus, quite rightly, on measures that help to prevent people from starting smoking, and help existing smokers to quit. However, once established, smoking is a very difficult addiction to break, and millions of people smoking today will never succeed. At present rates of progress it will take over two decades for the prevalence of smoking in the UK to halve from current levels, such that by 2025 there will probably still be over five million smokers in the UK. Preventing harm to the health of these smokers is a vital priority in this country, and in all countries where the smoking epidemic is established.

The Royal College of Physicians first called for radical policies to reduce the prevalence of smoking in 1962. Several of the policies we recommended then have since become established international practice. However, those measures, then and now, do not address the problem of smokers who cannot quit. The majority of the 150 million deaths from smoking expected worldwide in the next 20 years will occur in people who are smoking today. These people need help.

In this report we make the case for harm reduction strategies to protect smokers. We demonstrate that smokers smoke predominantly for nicotine, that nicotine itself is not especially hazardous, and that if nicotine could be provided in a form that is acceptable and effective as a cigarette substitute, millions of lives could be saved. We also argue that the regulatory systems that currently govern nicotine products in most countries, including the UK, actively discourage the development, marketing and promotion of significantly safer nicotine products to smokers.

Harm reduction is a fundamental component of many aspects of medicine and, indeed, everyday life, yet for some reason effective harm reduction principles have not been applied to tobacco smoking. This report makes the case for radical reform of the way that nicotine products are regulated and used in society. The ideas we present are controversial, and challenge many current and entrenched views in medicine and public health. They also have the potential to save millions of lives. They deserve serious consideration.

October 2007   **John Britton**
                          Chair, Tobacco Advisory Group of the Royal College of Physicians

1 | Use of tobacco in society

1.1 Introduction

An estimated 1.3 billion people worldwide currently smoke tobacco, mostly in the form of cigarettes.[1] The number of smokers is growing, particularly in middle- and low-income income countries where cigarettes are marketed aggressively by some of the world's most powerful companies, and is expected to reach 1.6 billion by 2025.[2] The health impacts of this inexorable spread in smoking are daunting. Smoking is already the leading cause of premature death in the developed world and is rapidly reaching that status in the developing world.[3]

Historically, tobacco has been used in a wider variety of forms, including smokeless tobacco. Use of smokeless tobacco has remained widespread in parts of the world and, in some countries, particularly Sweden and the United States, is once again increasing. This chapter reviews the history and diversity of tobacco use over time and place, examines the global spread of the cigarette epidemic, and reviews briefly the health impacts of different forms of tobacco use.

1.2 The history of tobacco use

1.2.1 Origins

Although historical accounts of tobacco use vary,[4–6] the fact that tobacco is considered so powerful in native American culture that it is thought to play a part in creation itself suggests that tobacco use is an ancient activity in the Americas.[7] The tobacco plant is indigenous to the American continent. By 15000 BC humans there may have begun to pick and use wild tobacco species, and smoked tobacco as part of ceremonial practices. By 5000 BC it is likely that tobacco cultivation began simultaneously with maize-based agriculture in central Mexico.[5] Thence, tobacco use and cultivation spread northwards – the first archaeological evidence for tobacco use, found in New Mexico, dates to 1400 BC – and smoking, originally limited to medicinal and ceremonial rituals, became adopted by wider society.[5,6]

By the time the Europeans arrived on the continent, tobacco use among indigenous Americans throughout North and South America was widespread and a core element of virtually all Native American cultures.[5,8] Christopher Columbus was presented with tobacco leaves on his arrival in October 1492 and he and his fellow explorers provide the first written documentation of tobacco use.[5] Records suggest that tobacco was being used in a variety of forms: it was smoked, chewed, inhaled as a powder, drunk as a tea, inserted as a liquid enema and consumed as a jelly. The natives of North America, for example, smoked tobacco with a pipe, while the Mayas mainly used crude cigars and cigarettes.[4,7]

1.2.2 The global spread of tobacco use begins

A few years after Columbus' arrival in America, tobacco leaf and seeds were brought back to Spain and Portugal. Tobacco use then gradually spread throughout Europe, Russia and the Middle East.[4,9] In 1560, Jean Nicot, the French ambassador to the Portuguese court (after whom nicotine was named), introduced tobacco and smoking to the French court as a medicine. Between 1530 and 1600, Spanish and Portuguese traders introduced tobacco to China, Japan, the Philippines, India and Africa.[10] Oceania was the last continent reached by tobacco – Captain Cook arrived smoking a pipe in 1769 and was promptly doused with water lest he be a demon.[10]

The predominant mode of tobacco use has changed considerably over time. In Europe, for example, pipe use predominated throughout the 17th century, to be replaced by snuff at the turn of the 18th century.[11] By the end of the 18th century, snuff taking was in decline and a revival in smoking, initially in the form of cigars, occurred, although largely among the wealthier classes.[8,11] Spain became a centre for cigar manufacture in the 1600s, and the European cigarette probably originated in the practice of beggars who rolled fragments of tobacco from used cigars in paper to make *papaletes*.[4] Philip Morris, which is now the world's largest privately run cigarette company, began life in 1847 as a tobacconist on Bond Street in London, selling hand-rolled Turkish cigarettes. The Crimean War (1853–6) popularised cigarette smoking among British soldiers who copied the art of hand-rolling tobacco from their Turkish allies. The first factories making hand-rolled cigarettes opened in the UK in the 1850s.

In the USA also, cigars, snuff and chewing tobacco were popular before cigarettes, which were first made from scraps left over after the production of other tobacco products. Cigarettes did not gain widespread popularity until after the American Civil War (1861–5). The origins of the US tobacco companies reflect this pattern. Lorillard was established in 1760 to process tobacco, cigars, and snuff. The Liggett & Myers Tobacco Company developed from a family snuff business established in the early 1800s. Twelve years after its official formation in 1873, it became the largest manufacturer of plug chewing tobacco in the world. The RJ Reynolds Tobacco Company was established in 1875 to produce chewing tobacco.

Elsewhere, the traditional form of tobacco use, until cigarettes were introduced, was often pipe smoking.[5] In Japan, although cigar smoking was introduced first via the Portuguese, smoking in long thin pipes (kiseru), probably introduced by the Dutch, became far more popular and came to accompany the tea ceremony.[5] Cigarettes were only introduced after the Meiji Restoration of 1868.[12] In China, tobacco pipe smoking was also traditional and, although initially seen as taboo (with various unsuccessful prohibitions ordered), it became so widely accepted by the 18th century that even women and children smoked.[5]

1.2.3 Mass production of cigarettes and the launch of the smoking epidemic

The nature and scale of tobacco use changed irrevocably in the early 1880s with the introduction of the Bonsack machine (a cigarette-rolling machine to replace hand rolling) and the mass manufacture of cigarettes.[11] Compared with traditional smoking methods, manufactured cigarettes were relatively clean, easy to use and increasingly affordable.[13] From that point on, cigarettes, and white cigarettes in particular, became the dominant form of tobacco used, and cigarette smoking spread globally on a massive scale.[2] The mass manufacture of cigarettes stimulated cigarette marketing, initially to overcome resistance to machine-made goods but subsequently to create demand for the vastly increased production.[14] Competition between the tobacco companies in the US and UK markets escalated, prompting the first series of tobacco company mergers and the launch of the global cigarette manufacturers, known today as the transnational tobacco companies.[15] The smoking epidemic had begun.

1.2.4 Continuing diversity in tobacco use

By the end of the 19th century, manufactured cigarettes and various types of hand-rolled cigarette accounted for up to 85% of all tobacco consumed worldwide.[2]

However, in contrast to the almost exclusive use of white cigarettes in the West, tobacco was, and continues to be, smoked in many different ways elsewhere. Bidis, which comprise tobacco hand-rolled in a non-tobacco leaf, are commonly used across much of South East Asia, and in India outsell cigarettes by seven to one. The Indonesian market is dominated by kreteks, cigarettes in which tobacco is blended with cloves, producing an anaesthetic agent (eugenol) which leads to deeper inhalation and high tar yields.[16] The water pipe (shisha, hookah or hubbly bubbly) is commonly used to smoke tobacco in many countries of North Africa, the Mediterranean and parts of Asia, and clay pipes (suipa, chilum or hookli) are used in South East Asia.[10,16] These and other methods of tobacco use are discussed in more detail in Chapter 5. Even manufactured cigarettes vary widely in design, for example in the use of filters, paper and additives. They also vary in the kind of tobacco used and how it is processed – factors that in turn influence the physiochemical nature of the smoke. In the United Kingdom, for example, flue-cured tobacco is preferred, while elsewhere in Europe a mixture of air-cured and flue-cured tobacco predominates.[17]

Traditional forms of smokeless tobacco use do persist, leading to considerable national and regional variation in tobacco habits. Smokeless tobacco use is widespread across much of South, South East and Central Asia, North Africa and the Middle East. In Mumbai, India, for example, 56% of women chew tobacco.[10] There is a wide variation in the type of tobacco used and in the ingredients added (see Box 1.1).[10,18,19]

In developed countries, use of smokeless tobacco is far less common. In the UK, for example, smokeless tobacco is used almost exclusively by minority groups of South Asian origin, among whom usage is thought to vary according to community and gender from 27% to 98%.[20] In the United States, an estimated 3% of adults use smokeless tobacco (including chewing tobacco and snuff) compared with about 20% who smoke cigarettes,[21,22] though use of smokeless tobacco has increased over recent years.[23]

An exception to the dominance of the cigarette in developed countries is Sweden where snus, a moist snuff that can be rolled by the user or bought in soft porous packs which are put under the upper lip, is used widely. Snus has been used in Sweden since 1637, and Sweden has the highest rate of per capita consumption of snuff in the world.[24] The high rates of snus use among Swedish men, combined with the fact that male smoking and tobacco-related mortality rates in Sweden are the lowest in the world and have been for some time,[25] has led to a growing interest in the role that less harmful smokeless tobacco products could play in smoking cessation and harm reduction as part of a broad approach to the provision and regulation of alternative sources of nicotine.[26,27] However,

---

Box 1.1 Forms of tobacco use.

Smoking tobacco

Manufactured cigarettes Machine-manufactured cigarettes containing shredded or reconstituted tobacco processed with hundreds of chemicals and wrapped in paper, usually with a filter.

Bidis Small, thin, hand-rolled cigarettes containing small volumes of tobacco hand-wrapped in non-tobacco leaves, usually dried temburni leaves. Common in South Asia, particularly India.

Kreteks Clove-flavoured cigarettes commonly consumed in Indonesia.

Cigars Air-cured and fermented tobaccos within a tobacco leaf wrapper. Vary in size from cigarette-sized cigarillos to much larger versions. Vary in form regionally, with some forms smoked with the ignited end placed inside the mouth.

Pipes Tobacco placed in the bowl of the pipe and inhaled through the stem, sometimes through water.

Smokeless tobacco

Often referred to as spit tobacco, spitting tobacco or swallow tobacco. Produced in two main forms – chewing tobacco and snuff:

Chewing tobacco Taken orally by placing a pinch of the mixture in the mouth between the gum and cheek and gently sucking and chewing. A wide variety of regional forms exists with use most common in Asia, the Middle East and the US (and migrant groups from these regions). Forms include: *gutkha, khaini, mawa, mishri, pan masala, qiwam, zarda* (Asia), *shammah* and *zarda* (Middle East), *loose-leaf, moist plug, plug and twist-roll* (US). The tobacco is often mixed with a variety of other compounds including spices or sweeteners, lime (calcium hydroxide) and other psychoactive substances such as areca nut.

Moist snuff Ground tobacco taken orally and held in the mouth between cheek and gum. Increasingly pre-packaged into small paper or cloth packets. Forms include *snus* (Sweden and Norway), *nass or naswa* (central Asia and the Middle East) and *toombak* (Sudan). A variety of brands are available in the US including *Skoal, Skoal Bandits* and *Timber Wolf*.

Dry snuff Powered tobacco inhaled through the nose or held between the lip and gum or cheek. Once popular, its use is now in decline.

Adapted from the *Tobacco atlas*,[10] Smokeless tobacco factsheets,[18] and CDC factsheet.[19]

the supply of snus or other oral tobacco products that are designed to be sucked was prohibited in the European Union in the early 1990s, so, with the exception of some minority use in Sweden's neighbouring countries, this product is not used appreciably elsewhere in Europe. Smokeless products that are designed to be chewed, including those used by the South Asian community, remain legal.

1.3 The smoking epidemic

1.3.1 A conceptual model

The onset of mass manufacture of cigarettes precipitated the smoking epidemic. A four-stage model of the epidemic has been described, based on observations of trends in cigarette consumption and tobacco-related diseases in western countries with the longest history of cigarette use, namely the United Kingdom and the United States (Fig 1.1; Table 1.1).[28] The model describes the typical sequence of uptake of smoking in men and women, and the subsequent occurrence of the harmful consequences of smoking. The epidemic begins with a rise in male smoking, followed by a period of stable high rates and then gradual decline. The onset of female smoking typically occurs later but then follows a similar pattern. Tobacco-related mortality among men and women follows the rise and fall of smoking prevalence, but with a lag of about 30–40 years.

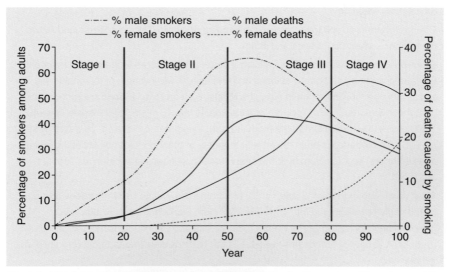

Fig 1.1   The four-stage evolution of the smoking epidemic. Reproduced from Lopez AD, Collishaw NE, Piha T. A descriptive model of the cigarette epidemic in developed countries. *Tob Control* 1994;3:242–7, with permission from the BMJ Publishing Group.[28]

The delay between the onset of smoking and its health impacts is a key feature of the epidemic, and one that explains why historically there has been confusion about the health impacts of tobacco. Smoking kills through causing a wide range of diseases, principally cancers, vascular disease and chronic lung disease (see Chapter 6 for further discussion). The delay between the onset of smoking and the occurrence of disease varies at individual and population level and from one disease to another. Death rates from lung cancer in a population do not reach

Table 1.1. The four stages of the smoking epidemic.

	Male		Female		Cigarette consumption (per capita)	Tobacco control
	Smoking prevalence	Deaths from tobacco	Smoking prevalence	Deaths from tobacco		
Stage 1 (1–2 decades)	Comparatively low <15%, but increasing.	At the start, deaths from tobacco similar to non-smoking populations. By the end of this phase some male deaths attributable to tobacco.	Rare: 0–10%.	As in non-smoking populations.	Low: <500/adult (most consumed by men).	Tobacco control underdeveloped. Smoking becomes socially acceptable.
Stage 2 (2–3 decades)	Continues to increase rapidly, peaks at 50–80%. Relatively few ex-smokers. Smoking rates similar in different social classes and may be higher in upper classes.	By the end of this phase ~10% male deaths due to tobacco. Male lung cancer rates increase approximately 10-fold from 5 to 50 per 100,000.	Increases rapidly, but lags behind men by 1–2 decades.	Rise in lung cancer rates much less. May reach 8–10 per 100,000.	1,000–3,000 (majority consumed by men). Men: 2,000–4,000.	Tobacco control not well developed.

continued over

Table 1.1. The four stages of the smoking epidemic – continued.

	Male		Female		Cigarette consumption (per capita)	Tobacco control
	Smoking prevalence	Deaths from tobacco	Smoking prevalence	Deaths from tobacco		
Stage 3 (3 decades)	Plateaus at a high level, then starts to decline. Lower in middle and older age men who give up. Male and female prevalence rates become more similar. Declines in prevalence greater among better educated.	Rapid rise in smoking-related mortality (to approximately 30% deaths) and higher in middle age. By the end of this phase lung cancer rates peak at around 110–120 per 100,000.	Plateaus at a lower level than in men (max 35–45%) and often for prolonged period, then declines late in this phase. Marked age gradient with higher rates in young women. Declines in prevalence greater among better educated.	Still relatively low (5%), but rising. By the end of this phase lung cancer rates reach approximately 23–40 per 100,000.	Men: 3,000–4,000. Women: 1,000–2,000.	Tobacco control improves. Smoking becomes less socially acceptable. Conditions for enacting tobacco control measures become more favourable.
Stage 4	Continues to decline but slowly. Male prevalence might be around 33–55%. Social class differences persist.	Peak in smoking-related mortality at 30–35% of all deaths (40–45% in middle age) during the early part of this stage. About one decade later, mortality from tobacco will start to decline.	Continues to decline but slowly. Social differences persist.	Rising rapidly. Will peak at ~20–25% of all deaths 2–3 decades into this phase (approximately 2 decades after the male peak) and decline thereafter.		Continued change in social climate. Smoke-free legislation becomes more feasible.

Adapted from Lopez et al and reproduced with permission from the BMJ Publishing Group Ltd.[28]

their maximum until 30–40 years after the peak in smoking prevalence, while deaths from vascular diseases will occur somewhat earlier.[28] Similarly, the excess risk of death from cardiovascular disease for an individual smoker declines quite quickly after quitting smoking, while the risk of lung cancer falls more slowly.[28,29] The delayed occurrence of health impacts means that countries in the early stages of the epidemic may be able to prevent much of the future growth in tobacco-related disease if effective public health interventions are implemented immediately. However, in countries with long-established tobacco epidemics, or where prevalence has been growing for several decades, tobacco-related mortality can only be reduced in the short term if large numbers of current smokers quit.[30] Reducing uptake among young people will only reduce mortality many decades into the future.

Other trends suggested by the model include the one-to-two decade delay between the increase in male and female smoking and the changing social pattern of smoking. Smoking is initially more common among the upper classes – uptake of any 'innovation' being usually more rapid among wealthier people. However, as the better off quit in response to health promotion messages, this pattern reverses, resulting in considerable social inequalities in smoking habits and health in the later stages of the epidemic. This is discussed briefly below and described further in Chapter 10.

1.3.2 The smoking epidemic in the United Kingdom

In the UK, the smoking epidemic took hold from the start of the 20th century. Male smoking increased rapidly over the next 50 years,[13] particularly during the two world wars, and by the late 1940s an estimated 65% of men were cigarette smokers. Male smoking prevalence then began to fall, and did so consistently until the early 1990s (Fig 1.2).[31] Women began to smoke approximately 20 years after their male counterparts, once social barriers to female smoking had begun to be dismantled. The tobacco industry played a key role in this process by promoting smoking as a symbol of emancipation, assisted by social changes during the world wars.[13] Female smoking then increased steadily, reaching a peak some 20 years later than male smoking in about the mid-1970s, and declined thereafter.

The UK is now in the later stages of the smoking epidemic model. Rates of smoking fell substantially through the 1970s and 1980s, particularly in men. Since the early 1990s, male and female smoking rates have been fairly similar. This pattern is typical of the late stages of the epidemic (though in some northern European countries female smoking is now more prevalent than male

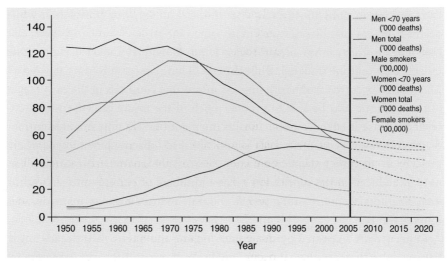

Fig 1.2 The UK smoking epidemic (projected data shown by dotted lines). Reproduced from West with permission.[31]

smoking). The rate of decline of smoking prevalence in the UK has since slowed considerably with little change through the 1990s.[32] Detailed analysis shows that a significant decline of about 0.4% per year occurred between 1999 and 2002,[33] and the latest data (from the 2004/5 General Household Survey) again suggest a small decline in prevalence. Of adults in Britain aged 16 or over, 25% (26% men and 23% women) now smoke cigarettes (Fig 1.2).[31,32]

Another key feature of this late stage of the epidemic is the social inequality in smoking and smoking-related health that emerges as smoking becomes increasingly entrenched among the least well off in society (see Chapter 10). In the UK, current smoking rates show an approximate two-fold variation between professional and working classes,[33–35] though these figures hide a far greater disparity between the most and least deprived groups.[36] These gradients are among the largest in Europe.[37] Other differences in smoking behaviour between social groups include poorer smokers starting to smoke at a younger age, consuming more tobacco, and finding it harder to quit.[36,38]

Mortality from tobacco, although still substantial (see Chapter 6), is also now declining in both men and women in the UK. Owing to the trends in smoking described above, the decline in male tobacco-related mortality occurred a few decades earlier than in women, in whom the decline has only recently begun.[36] Nevertheless, tobacco remains the leading cause of morbidity and mortality in the UK, recently estimated to cause 106,000 deaths each year, or about one in six of all deaths.[38–42] Moreover, owing to the social gradients in smoking habits described above, smoking is the leading cause of health inequalities, thought to account for half of the difference in survival to 70 years of age between the professional and unskilled manual social classes.[38]

1.3.3 The smoking epidemic elsewhere

The smoking epidemic in other northern and western European countries, in North America and in Australasia is also relatively advanced, but elsewhere is typically at an earlier stage. Some countries in sub-Saharan Africa are still at stage one of the epidemic, with low rates of cigarette smoking limited largely to men, and with little or no increase in tobacco-related diseases yet evident.[1] Some countries in Asia, North Africa and Latin America fit stage two of the epidemic, characterised by higher rates of male smoking and an increase in tobacco-related diseases in men beginning to occur, while smoking among women remains a relatively new phenomenon. Other countries in these regions are moving towards stage three of the epidemic – also typical of some countries in southern and eastern Europe – where male smoking rates are very high but are starting to fall, female smoking rates are close to reaching their peak, but tobacco-related deaths continue to rise in both genders.[1]

Not all countries fit the smoking epidemic model.[1] For example, in the countries of the former Soviet Union, tobacco smoking in men began around the end of the 19th century, at the same time as in the UK. However, the epidemic subsequently developed slightly differently, probably because cigarettes were produced and sold by a state monopoly rather than by the privately run transnational tobacco companies operating in the West. Thus, tobacco marketing was unknown, and smoking among women remained socially less acceptable and far less common. However, since the collapse of the Soviet system in the early 1990s, smoking rates in women have increased substantially; rates in women in Russia, for example, doubled between 1992 and 2003.[43] In men, the very high rates of smoking (between 50% and 65%)[44] have failed to decline as the model would predict. Instead, increases have been observed in Russia, the only former Soviet country in which longitudinal data have been analysed.[43] These increases can be attributed to the aggressive marketing by the transnational tobacco companies as they gained access to markets in the region and the absence of effective tobacco control measures.[44]

1.3.4 The vector of the epidemic

Such patterns in the development of tobacco use within populations provide evidence of how the tobacco industry, particularly transnational tobacco companies, acts as the vector for the epidemic. Patterns of smoking uptake can be explained largely by the presence and activity of the transnational tobacco companies, particularly from the 1950s onwards, when tobacco companies began to increase their international investments. During the 1970s, major investments were made in Latin America, in the 1980s in Asia, and in the 1990s in the former Eastern bloc countries.[45,46,47] China, the world's largest tobacco market, remains the primary target for expansion, with companies jostling to establish the first substantial joint venture with the China National Tobacco Corporation.[48]

This global expansion was first triggered by health scares in the West,[14,49,50] which resulted in declining tobacco consumption, and subsequently by global political and economic change, including trade and investment liberalisation, and the opening of formerly closed markets. Trade and investment liberalisation has contributed significantly to increases in cigarette consumption, particularly in low- and middle-income countries.[2,51] It is also notable that the most marked increases in smoking prevalence have occurred in markets newly targeted by the transnational tobacco companies, and particularly among women. As a result, and contrary to the view that could easily be formed by an affluent citizen in a developed country, the global smoking epidemic continues to grow. The consequences of this global expansion are startling. It has caused global cigarette consumption to continue to increase. Falling per capita consumption in industrialised countries (by 10% between 1970 and the mid-1990s) has been more than offset by rises in low- and middle-income countries (Fig 1.3).[2,52] Over three quarters of the world's smokers (most recently estimated at 1.3 billion smokers) now live in low- and middle-income countries.[1,53] While an estimated 35% of men in developed countries smoke, this compares with almost 50% of men in developing countries and almost two thirds of Chinese men.[1] As a result, the future global health burden from tobacco will continue to increase, particularly in women, and will shift sharply in geographical focus from high- to low- and middle-income countries as outlined further below.

1.4 The health impacts of tobacco smoking

Smoking is uniquely harmful: half of all long-term smokers will eventually be killed by their habit and, of these, half will die during middle age, losing 20–25 years of life.[3,54,55] Smoking has now been positively associated with over 40 diseases and the list continues to grow (see Chapter 6).[56] For most diseases, the association with smoking is strong and viewed as causal. Specifically, the major associations between smoking and disease satisfy the commonly accepted criteria for causality. Thus, the associations are strong and consistent (having

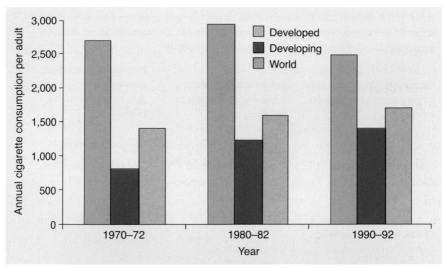

Fig 1.3 Trends in per capita adult cigarette consumption. Reproduced with permission from the International Bank for Reconstruction and Development/The World Bank.[2]

been found in numerous studies, using different designs, and in different populations), with evidence of a relation. The temporal relation between smoking and disease has been established and plausible biological mechanisms exist. The diseases fall into three main categories: cancer, vascular diseases and chronic lung diseases (Table 1.2).[54,56-61]

Most of the evidence on the health impacts of smoking comes from high-income countries but more recent evidence is emerging from other countries including China and India.[62-65] These studies suggest that the overall risks of smoking are about as great as in high-income countries and the diseases caused by smoking are similar. However, the specific patterns of smoking-related diseases may differ, reflecting the pre-existing distribution of underlying causes of death. In China, compared with the West, tobacco causes far more deaths from chronic lung diseases than from vascular disease. Smoking kills by making smoking-related diseases that are already present in the population much more common.

The components of cigarette smoke collectively explain its health impacts.[66,67] Smoke contains over 4,000 chemicals, approximately 500 in the vapour phase (including carbon monoxide, ammonia, nitrogen oxides, hydrogen cyanide, and various hydrocarbons) and over 3,500 in the particulate phase (including 'tars' and most of the carcinogenic agents). Nicotine appears in both phases.[17] Nicotine is the predominant addictive chemical and the reason why smokers continue to smoke (see Chapters 2, 3 and 4). However, cigarette smoke also contains carcinogenic agents in the particulate phase, including tobacco-specific

Table 1.2. Relative risks in current male smokers versus males who have never smoked (female data for cervical cancer) based on UK Doctors Study and the American Cancer Society Cancer Prevention Study II.

Increased risk in smokers largely or entirely due to smoking	Increased risk among smokers partly due to smoking	Increased risk among smokers largely or partly due to confounding factors
Cancers		
Mouth, pharynx, larynx 24, 24.5	Oesophagus 7.5, 7.6	Liver 1.6
Lung 15, 22.4	Leukaemia (myeloid) 1.8,	Cervix 2.1
Bladder 2.3, 2.9	Kidney 2.1, 3	Large bowel
Pancreas 2.2, 2.1	Stomach 1.7	
Cardiovascular disease		
Ischaemic heart disease 1.6, 1.9	Cerebral vascular disease 1.5, 1.9	
Hypertension 1.4, 2.4		
Other heart disease 2.1		
Aortic aneurysm 4.1, 6.3		
Peripheral vascular disease 9.7		
Respiratory diseases		
Chronic obstructive pulmonary disease 12.7, 17.6	Pneumonia 1.9	
Asthma 2.2, 1.3		
Other diseases	Peptic ulcer	Cirrhosis
	Crohn's disease	Suicide
	Osteoporosis	Poisoning
	Periodontitis	
	Tobacco amblyobia	
	Age-related macular degeneration	
	Cataracts	
	Hip fracture	

Adapted from data in Doll,[56] Boyle,[57] Doll,[60] and Wald and Hackshaw.[61]

N-nitrosamines (TSNAs), polynuclear aromatic hydrocarbons (PAH), and tobacco non-specific N-nitrosamines.[17] The carbon monoxide, nitrogen oxides and other gaseous constituents of cigarette smoke have been shown to play a particular role in cardiovascular disease: they reduce oxygen transport, promote the atherosclerotic process, alter the serum lipid profile and increase platelet stickiness and tendency to aggregate.[68]

The exact composition of tobacco smoke is primarily determined by the type of tobacco (including how it has been manufactured, cured and stored), although variations in cigarette design, including filtration, ventilation, paper and additives and the manner in which individuals smoke, also has an influence (see Chapter 5).[17,67]

1.4.1 Impacts of smoking on global health status

The impacts of smoking on global health status are daunting. Smoking is the single largest cause of avoidable, premature death in economically developed countries, and is rapidly achieving this status outside the developed world.[28] Peto *et al* estimate that in 1995 smoking caused three million deaths globally and that by approximately 2025 this will increase to about 10 million deaths each year. Moreover, the global distribution of these deaths will change dramatically, reflecting the shift in the epidemic described above. In 1995, two thirds of deaths caused by tobacco occurred in the developed world and accounted for one in six deaths there; by 2025, three quarters of deaths caused by tobacco will be in the developing world.[3,52]

1.4.2 The health impacts of smoking in context

Since the early 1990s, efforts have been made to assess the comparative importance of different risk factors on health, both globally and in regional population groups, in order to help determine policy and research priorities. Recently published analyses based on 2001 data compare smoking with 19 other risk factors.[69] These show that in high-income countries smoking is the leading cause of death, responsible for 18.5% (almost one in five) of all deaths. In low- and middle-income countries, estimates suggest that only high blood pressure and childhood underweight currently cause more deaths than smoking, and that smoking is responsible for 6.9% of the total (more than one in 15 deaths). Globally, smoking is estimated to have caused 4.8 million deaths in 2001; 8.5% of the total (approximately one in 12 deaths). This is in line with the estimates given above confirming the upward trajectory of the tobacco epidemic.

Tobacco also causes considerable long-term morbidity. Disability adjusted life years (DALYs) are a measure of healthy life lost as a result of individuals being in a state of poor health or disability. One DALY can be thought of as one lost year of healthy life. Estimates have been made using DALYs to quantify the impact of these risk factors on morbidity. Smoking was identified as the leading cause of disability in high-income countries, where it is responsible for 13% of DALYs lost. In low- and middle-income countries it is responsible for 4% of DALYs lost (Fig 1.4).[69] Globally, smoking in 2001 was responsible for the loss of 72.9 million DALYs, almost 5% of the total. Along with mortality, the global burden of morbidity from tobacco is predicted to increase.[69,70] The 2001 figures are almost double the estimated DALYs lost globally to tobacco use in 1990. A further rise to over 120 million DALYs lost to tobacco – just over 9% of the global total – is predicted by 2020.[70]

1.4.3 The health impacts of other forms of tobacco use

Cigars, pipes and bidis

The data given above relate almost exclusively to cigarette smoking. The health consequences of smoking tobacco in other forms have been less thoroughly evaluated, but are thought to be broadly similar. The smoke from cigars, pipes and bidis has been shown to be carcinogenic and studies show an increased risk of cancers of the oral cavity, pharynx, larynx, oesophagus, lung and stomach – a subset of those already shown to be linked with cigarette smoking.[71] Compared with cigarette smokers, cigar smokers appear to have similar risks of oral and oesophageal cancers, but lower risks of lung and laryngeal cancer, coronary heart disease and chronic obstructive pulmonary disease, probably because cigar smokers tend not to inhale and smoke less tobacco in total.[72] Those who switch from cigarette smoking to cigar or pipe smoking reduce their risk of death over those that continue to smoke cigarettes, but have considerably higher risks of dying than those who quit altogether.[72,73]

The changing cigarette

The form of the commercial cigarette has changed considerably since the 1950s, largely in response to health concerns. Filterless cigarettes have given way to filtered versions (virtually all cigarettes sold commercially in the UK are now filtered), the type and preparation of tobacco used in cigarettes has changed, various ingredients have been added, and machine-measured tar and nicotine yields (but not the actual tar and nicotine content or the tar and nicotine delivered in the process of normal smoking) have fallen considerably.[17,67,74] Some studies suggest that the introduction of filters in the 1950s and initial reduction in tar yields from very high levels (over 25 mg) may have had some health benefits, particularly in reducing some cancer risks, but there has been little or no apparent effect on risk of vascular or chronic obstructive pulmonary disease.[57,68,75] Any potential benefits have, however, largely been swamped by the introduction and

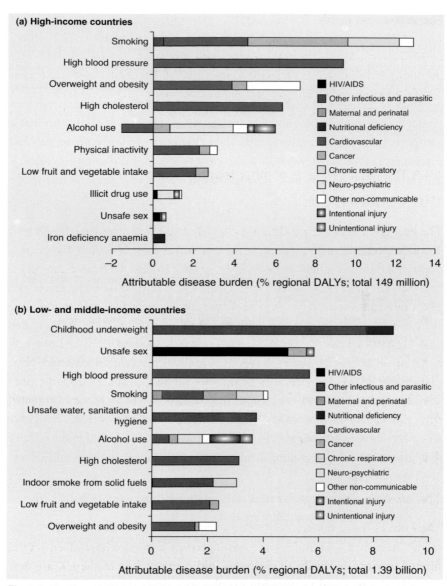

Fig 1.4 Burden of disease attributable in 2001 to 10 leading regional risk factors, by disease type in (a) high-income countries and (b) low- and middle-income countries. Reproduced from Lopez *et al* with permission of the World Bank.[69]

marketing of so called 'light' cigarettes – a cynical attempt by the industry to encourage smokers concerned about health risks to switch to 'lights' rather than quit. Evidence suggests that although such cigarettes may have lower machine-measured nicotine and tar yields, the actual tar and nicotine delivered during smoking differs little from 'full flavour' cigarettes (see Chapter 5).[76]

Smokeless products

The health risks of smokeless tobacco are considerably lower than those associated with combustible tobacco products as it is largely the combustion process that makes tobacco use so deadly.[77,27] The evidence on the health effects of smokeless products is reviewed in Chapter 8. The major benefits of smokeless tobacco products over combustible products are the virtual absence of respiratory risks, and potentially lower cardiovascular risks due to the absence of the gaseous components in smoke. However, it is also clear that the risks of smokeless tobacco use vary between products and, although low in relation to cigarette smoking, are not inconsequential.

1.5 Conclusions

▶ Tobacco use originated in the American continent thousands of years ago and has spread to the rest of the world in the last 500 years.

▶ Tobacco continues to be used in many different ways, but the most common form of consumption is now the cigarette.

▶ The current global epidemic of cigarette smoking is a recent phenomenon, dating from around the turn of the 20th century.

▶ The onset of the smoking epidemic typically occurs in men before women, with epidemic increases in deaths caused by smoking occurring 20–30 years after the onset of smoking.

▶ In some developed countries, both smoking prevalence and mortality rates are now falling.

▶ Most countries are at an earlier stage of the epidemic, and globally both smoking rates and mortality are rising.

▶ Smoking is the biggest avoidable cause of premature death and disability in most developed countries, and with the evolution of the global smoking epidemic will be equally important in the future wherever smoking becomes prevalent.

▶ In 2001, smoking caused 4.8 million deaths, equivalent to about one in every 12 of all deaths, globally.

▶ By 2025, there will be an estimated 1.6 billion smokers in the world, and smoking will cause approximately 10 million deaths each year.

▶ Most of these deaths will occur in people who already smoke, rather than those who start smoking between now and 2025. Therefore, while preventing the uptake of smoking is crucially important to the prevention of deaths in the longer term, promoting smoking cessation has a greater effect on mortality in the shorter term.

▶ It is, therefore, crucial to find ways of helping existing smokers to quit smoking, as well as preventing the uptake of smoking.

References

1   Shafey O, Dolwick S, Guindon GE. *Tobacco control country profiles*, 2nd edn. Atlanta: American Cancer Society, 2003.

2   World Bank. *Curbing the epidemic: governments and economics of tobacco control.* Washington DC: World Bank, 1999. www.worldbank.org/tobacco/chapter1.asp (accessed 4 July 2007).

3   Peto R, Lopez A, Boreham J, Thun M, Clark HJ. *Mortality from smoking in developed countries 1950–2000: indirect estimates from national vital statistics.* Oxford: Oxford University Press, 1994.

4   Borio G. *Tobacco timeline* 2001. www.tobacco.org/resources/history/tobacco_history.html (accessed 4 July 2007).

5   Winter D. Native Americans. In: Goodman J (ed), *Tobacco in history and culture: an encyclopedia.* Farmington Hills MI: Charles Scribner's Sons, 2004.

6   Gilman SL, Xun Z. Introduction. In: Gilman SL, Xun Z (eds). *Smoke: a global history of smoking.* London: Reaktion Books, 2004.

7   Robicsek F. Ritual smoking in Central America. In: Gilman SL, Xun Z (eds), *Smoke: a global history of smoking.* London: Reaktion Books, 2004.

8   Tobacco smoking in Britain: an overview. In: *Nicotine addiction in Britain.* Report of the Tobacco Advisory Group of the Royal College of Physicians. London: RCP, 2000.

9   Doll R. Uncovering the effects of smoking: historical perspective. *Stat Methods Med Res* 1998;7:87–117.

10 Mackay J, Eriksen M. *The tobacco atlas.* Geneva: World Health Organization, 2002. www.who.int/tobacco/statistics/tobacco_atlas/en/ (accessed 4 July 2007).

11 Hilton M. *Smoking in British popular culture 1800–2000.* Manchester: Manchester University Press, 2000.

12 Feldman E. The limits of tolerance. In: Feldman E, Bayer R (eds), *Unfiltered: conflicts over tobacco policy and public health.* Cambridge MA: Harvard University Press, 2004.

13 Amos A, Haglund M. From social taboo to 'torch of freedom': marketing cigarettes to women. *Tob Control* 2000;9:3–8.

14 Shepherd PL. Transnational corporations and the international cigarette industry. In Newfarmer RS (ed), *Profits, progress and poverty: case studies of international industries in Latin America.* Notre Dame IN: University of Notre Dame Press, 1985.

15 Cox H. *The global cigarette: origins and evolution of British American tobacco 1990–1945.* Oxford: Oxford University Press, 2000.

16 Collin J, Gilmore A. Developing countries. In: Goodman J (ed), *Tobacco in history and culture: an encyclopedia.* Farmington Hills MI: Charles Scribner's Sons, 2004.

17 Hoffman D, Hoffmann I. The changing cigarette 1950–1995. *J Toxicol Environ Health* 1997;50:307–64.

18 Smokeless Tobacco Factsheets. *3rd International Conference on Smokeless Tobacco.* Stockholm, Sweden: September 2002.

19   Centers for Disease Control and Prevention. *Betel quid with tobacco (gutka)*. Factsheet, May 2006. www.cdc.gov/tobacco/factsheets/BetelQuidTobacco_factsheet.htm (accessed 7 August 2007).

20   McNeill A, Bedi R, Islam S *et al.* Levels of toxins in oral tobacco in the UK. *Tob Control* 2006;15:64–7.

21   Centers for Disease Control and Prevention. S*mokeless tobacco*. Factsheet, April 2007. www.cdc.gov/tobacco/factsheets/smokelesstobacco.htm (accessed 4 July 2007).

22   Centers for Disease Control and Prevention. *Adult cigarette smoking in the United States: current estimates*. Factsheet, November 2006. www.cdc.gov/tobacco/factsheets/ AdultCigaretteSmoking_FactSheet.htm (accessed 4 July 2007).

23   Pershagen G. Smokeless tobacco. *Br Med Bull* 1996;52:50–7.

24   Idris AM, Ibrahim SO, Vasstrand EN. The Swedish snus and the Sudanese toombak: are they different? *Oral Oncol* 1998;34:558–66.

25   Rodu B, Stegmayr B, Nasic S, Asplund K. Impact of smokeless tobacco use on smoking in northern Sweden. *J Intern Med* 2002;252:398–404.

26   Bates C, Fagerstrom K, Jarvis MJ *et al.* European Union policy on smokeless tobacco: a statement in favour of evidence based regulation for public health. *Tob Control* 2003;12:360–7.

27   Foulds J, Ramstrom L, Burke M, Fagerstrom K. Effect of smokeless tobacco (snus) on smoking and public health in Sweden. *Tob Control* 2003;12:349–59.

28   Lopez AD, Collishaw NE, Piha T. A descriptive model of the cigarette epidemic in developed countries. *Tob Control* 1994;3:242–7.

29   US Department of Health and Human Services. *The health benefits of smoking cessation*. Report of the Surgeon General. Rockville, MD: US Department of Health and Human Services, Public Health Service, Centers for Disease Control, Office on Smoking and Health, 1990.

30   Peto R, Darby S, Deo H *et al.* Smoking, smoking cessation, and lung cancer in the UK since 1950: combination of national statistics with two case-control studies. *BMJ* 2000;321:323–9.

31   West R. *Smoking: prevalence, mortality and cessation in Great Britain, 2005*. www.rjwest.co.uk/resources/smokingcessation.doc (accessed 7 August 2007).

32   Office for National Statistics. Cigarette smoking: slight fall in smoking prevalence. *General Household Survey, 2005*. London: The Stationery Office, 2006. www.statistics.gov.uk/cci/nugget.asp?id=866 (accessed 7 August 2007).

33   Jarvis MJ. Monitoring cigarette smoking prevalence in Britain in a timely fashion. *Addiction* 2003;98:1569–74.

34   Lader D, Goddard E. *Smoking-related behaviour and attitudes, 2004*. London: Office for National Statistics, 2005.

35   Office for National Statistics. *Prevalence of cigarette smoking by sex and socio-economic classification based on the current or last job of the household reference person: living in Britain*. www.statistics.gov.uk/STATBASE/ssdataset.asp?vlnk=5453 (accessed 4 July 2007).

36   Action on Smoking and Health. *Smoking and health inequalities*. Factsheet, November 2005. www.ash.org.uk/html/factsheets/html/healthinequalities2005/healthinequalities 05.doc (accessed 4 July 2007).

37   Kunst A, Giskes K, Mackenback J. *Socio-economic inequalities in smoking in the European Union: applying an equity lens to tobacco control policies*. Rotterdam: Erasmus Medical Center, 2004.

38   Jarvis MJ, Wardle J. Social patterning of health behaviours: the case of cigarette smoking. In: Marmot M, Wilkinson R (eds), *Social determinants of health*, 2nd edn. Oxford: Oxford University Press, 2005.

39   Peto R, Lopez A, Boreham J, Thun M. *Mortality from smoking in developed countries 1950–2000*, 2nd edn, revised June 2006. www.ctsu.ox.ac.uk/~tobacco/C4308.pdf (accessed 4 July 2007).

40   Twigg L, Moon G, Walker S. *The smoking epidemic in England*. London: Health Development Agency, 2004.

41   Health Check. *On the state of the public health*. Annual report of the Chief Medical Officer, 2002.

42   Wanless D. *Securing good health for the whole population: final report*. London: The Stationery Office, 2004.

43   Perlman F, Bobak M, Gilmore A, McKee M. Trends in the prevalence of smoking in Russia during the transition to a market economy. *Tob Control* (submitted).

44   Gilmore A, Pomerleau F, McKee M. Prevalence of smoking in 8 countries of the former Soviet Union: results from the living conditions, lifestyles and health study. *Am J Public Health* 2004;94:2177–87.

45   Connolly GN. Worldwide expansion of transnational tobacco industry. *J Natl Cancer Inst* 1992; monographs: 29–35.

46   Gilmore A, McKee M. Moving east: how the transnational tobacco companies gained entry to the emerging markets of the former Soviet Union. Part I: establishing cigarette imports. *Tob Control* 2004;13:143–50.

47   Gilmore A, McKee M. Moving east: how the transnational tobacco companies gained entry to the emerging markets of the former Soviet Union. Part II: an overview of priorities and tactics used to establish a manufacturing presence. *Tob Control* 2004;13: 151–60.

48   Lee K, Gilmore A, Collin J. Breaking and re-entering: British American Tobacco in China 1979–2000. *Tob Control* 2004;13(Supp 2):88–95.

49   Slade J. The tobacco epidemic: lessons from history. *J Psychoactive Drugs* 1989;21:281–91.

50   Council on Scientific Affairs. The worldwide smoking epidemic. Tobacco trade, use, and control. *JAMA* 1990;263:3312–8.

51   Gilmore A, McKee M. Exploring the impact of foreign direct investment on tobacco consumption in the former Soviet Union. *Tob Control* 2005;14:13–21.

52   Gajalakshmi CK, Jha P, Ranson S, Nguyen S. Global patterns of smoking and smoking-attributable mortality. In: Jha P, Chaloupka F (eds), *Tobacco control in developing countries*. Oxford: Oxford University Press, 2000.

53   Guindon GE, Boisclair D. *Past, current, and future trends in tobacco use*. The World Bank, February 2003. www.worldbank.org/tobacco/pdf/Guindon-Past,%20current-%20whole.pdf (accessed 7 August 2007).

54   Doll R, Peto R, Wheatley K, Gray R, Sutherland I. Mortality in relation to smoking: 40 years' observations on male British doctors. *BMJ* 1994;309:901–11.

55   Peto R, Lopez AD, Boreham J, Thun M, Heath C. Mortality from smoking in developed countries: indirect estimation from national vital statistics. *Lancet* 1992;339:1268–78.

56   Doll R. Risk from tobacco and potentials for health gain. *Int J Tuberc Lung Dis* 1999; 3:90–9.

57   Boyle P. Cancer, cigarette smoking and premature death in Europe: a review including the recommendations of European Cancer Experts Consensus Meeting, Helsinki, October 1996. *Lung Cancer* 1997;17:1–60.

58   US Department of Health and Human Services. *Reducing the health consequences of smoking: 25 years of progress.* Report of the Surgeon General. Rockville, MD: US Department of Health and Human Services, Public Health Service, Centers for Disease Control, Office on Smoking and Health, 1989.

59   Thun MJ, Day-Lally CA, Calle EE, Flanders WD, Heath CW. Excess mortality among cigarette smokers: changes in a 20-year interval. *Am J Pub Health* 1995;85:1223–30.

60   Doll R. Cancers weakly related to smoking. *Br Med Bull* 1996;52:35–49.

61   Wald NJ, Hackshaw AK. Cigarette smoking: an epidemiological overview. *Br Med Bull* 1996;52:3–11.

62   Gupta PC, Mehta HC. Cohort study of all-cause mortality among tobacco users in Mumbai, India. *Bull World Health Organ* 2000;78:877–83.

63   Chen ZM, Xu Z, Collins R, Li WX, Peto R. Early health effects of the emerging tobacco epidemic in China. A 16-year prospective study. *JAMA* 1997;278:1500–4.

64   Niu SR, Yang GH, Chen ZM *et al.* Emerging tobacco hazards in China: 2. Early mortality results from a prospective study. *BMJ* 1998;317:1423–4.

65   Liu BQ, Peto R, Chen ZM *et al.* Emerging tobacco hazards in China: retrospective proportional mortality study of one million deaths. *BMJ* 1998;317:1411–22.

66   Royal College of Physicians. *Health or smoking?* Follow-up report of the Royal College of Physicans. London: Pitman Publishing, 1998.

67   WHO International Agency for Research on Cancer. *Tobacco smoking.* IARC monograph on the evaluation of the carcinogenic risk of chemicals to humans. Lyon: IARC, 1986.

68   WHO International Agency for Research on Cancer. *Tobacco: a major international health hazard.* IARC Scientific Publication No. 74. Lyon: IARC, 1986.

69   Lopez AD, Mathers CD, Ezzati M, Jamison DT, Murray CJL. *Global burden of disease and risk factors.* Washington DC: the World Bank Group and New York: Oxford University Press, 2006. www.dcp2.org/pubs/GBD (accessed 7 August 2007).

70   Murray C, Lopez A. Global mortality, disability and the contribution of risk factors: global burden of disease study. *Lancet* 1997;349:1436–42.

71   WHO International Agency for Research on Cancer. *Volume 83: Tobacco smoke and involuntary smoking.* IARC monograph on the evaluation of the carcinogenic risk of chemicals to humans. Lyon: IARC, 2004. monographs.iarc.fr/ENG/Monographs/vol83/volume83.pdf (accessed 7 August 2007)

72   National Cancer Institute. *Cigars: health effects and trends.* Smoking and Tobacco Control Monograph No. 9. Bethesda MD: US Department of Health and Human Services, Public Health Service, National Institutes of Health, NIH Publication No. 98-4302, February 1998.

73   Wald NJ, Watt HC. Prospective study of effect of switching from cigarettes to pipes or cigars on mortality from three smoking related diseases. *BMJ* 1997;314:1860.

74   Royal College of Physicians. *Nicotine addiction in Britain.* Report of the Tobacco Advisory Group of the Royal College of Physicians. London: RCP, 2000.

75   Wynder EL, Stellman SD. Impact of long-term filter cigarette usage on lung and larynx cancer risk: a case-control study. *J Natl Cancer Inst* 1979;62:471–7.

76   National Cancer Institute. *Risks associated with smoking cigarettes with low machine measured yields of tar and nicotine.* Smoking and Tobacco Control Monograph No. 13. Bethseda, MD: US Department of Health and Human Services, Public Health Service, National Institutes for Health, NIH Publication No. 02-5074, October 2001.

77   Cnattingius S, Galanti R, Grafstrom R *et al. Halsorisker med svenskt snus.* Stockholm: Karolinska Institutet, November 2005.

2 | Nicotine and nicotinic receptors, and their role in smoking

2.1 Chemical and pharmacokinetic aspects of nicotine

Nicotine belongs to a large family of amine-containing chemicals called alkaloids. Most alkaloids are produced by plants. Although nicotine is particularly abundant in tobacco, detectable amounts are also found in some related plants, such as potato, aubergine and tomato. Nicotine in plants probably functions as an insecticide, and concentrated solutions of nicotine were once widely sold for this purpose, giving way in recent years to organophosphates.

The chemical name for nicotine is (S)-3-(1-Methyl-2-pyrroli-dinyl)pyridine, reflecting the existence of two nitrogen-containing carbon rings (Fig 2.1). In its pure form, nicotine is a colourless or pale yellow oily liquid. In chemical terms, pure nicotine is a base and is often referred to as nicotine free-base. It combines with acids to form stable, powdered salts which can be readily dissolved in water; this is the form of nicotine most commonly used by research scientists.

The absorption, distribution and fate of nicotine in the body has been studied extensively and reviewed recently.[1] Nicotine can be absorbed slowly or rapidly, depending on the route of administration and on the way the drug is formulated. Nicotine is most rapidly taken up after cigarette smoke inhalation, with arterial levels peaking approximately 20 seconds after each puff.[2] Although the kinetics

23

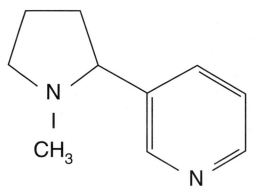

Fig 2.1 Chemical structure of nicotine.

of delivery of inhaled nicotine to the brain are still not fully understood, it is estimated that nicotine from smoked tobacco reaches the brain about 10 seconds faster than from the intravenously injected drug.[2] In contrast, the various forms of nicotine replacement therapy (patch, gum, nasal spray, inhaler etc) deliver nicotine much more slowly.[1]

In body fluids such as blood, most nicotine molecules are positively charged and will not pass readily through cell membranes. This is because nicotine is a weak base, with a pKa of about 8.0.[1] Nevertheless, a significant proportion (approximately 30%) of nicotine in the circulation is uncharged, and in this form it distributes readily to tissues including the brain. The brain efficiently extracts nicotine from the circulation and releases it back to the blood over the course of many minutes.[3,4] In abstinent smokers, nicotine levels in the blood initially decline with a half-life of about two hours. However, after a few hours of abstinence, nicotine disappears more slowly as the circulation continues to be replenished by nicotine stored in tissues.[1] Nicotine blood levels in smokers vary during the day,[5] typically rising during the morning and reaching a plateau in the early afternoon which continues until late evening. During the night, nicotine levels decline to about one third of their daytime maximum. A pattern of peaks and troughs corresponding to individual cigarettes is superimposed on this circadian rhythm.

Nicotine undergoes extensive metabolism in the body, primarily in the liver, and six primary metabolites have been recognised in humans.[1] Cotinine, the most abundant metabolite, has a much longer plasma half-life than nicotine (16–20 hours).[6] It is used in smoking cessation studies as a measure of tobacco smoke exposure and smoking to confirm self-reported abstinence. Animal experiments suggest that chronic nicotine exposure leads to an accumulation of cotinine and other metabolites in the brain.[7] This is potentially significant because some metabolites are known to be pharmacologically active.[1,8] For example, both cotinine and nornicotine (another major metabolite) are able to enhance the *in vitro* release of dopamine in animal studies,[9,10] and nornicotine is voluntarily self-administered by rats.[11] It is not known whether cotinine and nornicotine reach pharmacologically significant concentrations in the brains of smokers, but on present evidence it would certainly be premature to rule out a role for nicotine metabolites in the motivation to smoke. Any nicotine that is not metabolised is excreted, mainly by the kidney.

2.2 The physiological functions of nicotinic receptors

Nicotine produces pharmacological actions by targeting proteins whose natural role is to act as receptors for the neurotransmitter acetylcholine (ACh) (Fig 2.2). These receptors are commonly referred to as nicotinic cholinergic receptors or nAChRs. Acetylcholine also acts on a second type of receptor that is insensitive to nicotine but sensitive to a mushroom-derived drug, muscarine (termed muscarinic cholinergic receptors). Both ACh and nicotine activate nicotinic receptors, and they are therefore termed nicotinic agonists. Drugs which block these receptors are known as nicotinic antagonists.

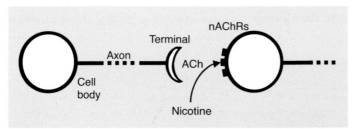

Fig 2.2 A cholinergic neuron (left) forming a synapse with a second neuron (right). The neurotransmitter acetylcholine (ACh) is released from a nerve terminal of the former and crosses the synapse to stimulate nicotinic receptors (nAChRs) located postsynaptically on the latter neuron. Nicotine, which is not naturally produced in the body, can gain access to the same nicotinic receptors as ACh. Once nicotine has bound to these receptors, it may stimulate or desensitise them, depending in part on the drug concentration and how long the drug is present. ACh = acetylcholine.

Acetylcholine performs a plethora of functions in the peripheral and central nervous systems, some mediated by nicotinic receptors and others by muscarinic receptors. The molecular structure of nicotinic receptors favours rapid communication between neurons. Thus, when nAChRs are activated by ACh or nicotine, they transmit a signal to the inside of the cell within milliseconds. In contrast, muscarinic ACh receptors tend to react within seconds to minutes. Rapid neurotransmission is obviously important at the junction between motor nerve terminals and skeletal muscles, and it is therefore unsurprising that nAChRs mediate this function. Nicotinic receptors also play a key role in autonomic nervous system activity, thereby modulating processes as diverse as digestion, hormone secretion, blood pressure control, and pupil constriction.

In the brain and spinal cord, only a very small percentage of neurons use ACh as a neurotransmitter, yet these neurons ramify widely, so that virtually the entire central nervous system is supplied with ACh-releasing terminals. Nicotinic and muscarinic receptors are also widely distributed. These two receptor types are sometimes found at the same synapses, but can also occur independently. In anatomical terms, the brain can be subdivided into regions (cerebral cortex, hypothalamus, striatum etc) and into several hundred smaller structures known as nuclei. Most, if not all, brain nuclei express nicotinic receptors.[12,13] Therefore, these anatomical considerations suggest that nicotine could exert a widespread influence on the brain through actions on nicotinic cholinergic receptors.

Nicotinic receptors are located on the outer surface of neurons and help to regulate their electrical activity. They do this principally by controlling the entry of electrically charged chemicals (ions, particularly sodium and calcium) into cells. Each nicotinic receptor is cylindrical in shape, and has a central channel through which ions can be permitted to pass into the cell (Fig 2.3). In the absence of an agonist, the nicotinic receptor assumes a 'resting' shape in which

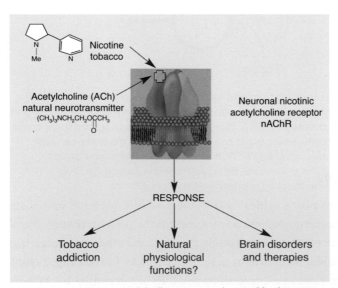

Fig 2.3 A nicotinic acetylcholine receptor located in the outer membrane of a neuron. The receptor comprises five subunits (shown in blue) surrounding a central channel, which can open to allow the passage of positively-charged ions (Na^+ and Ca^{++}) into the cell.

the ion channel is closed. When a nicotinic agonist is administered, some agonist molecules bind to the receptor and induce the receptor to change its shape so that the ion channel opens. The resultant flow of positively charged ions into the cell can have several consequences. First, the neuron becomes more electrically active, producing electrical impulses (action potentials) at a faster rate. When these impulses are conducted along the axon of the neuron to the nerve terminals, they result in increased release of the neurotransmitter. Second, some nicotinic receptors are capable of conducting calcium ions (Ca^{++}) into the neuron. Ca^{++} ions play a critical role as signalling molecules within the cell. By increasing Ca^{++} ion levels within the cell, nicotinic receptors can have a pronounced effect on many cell processes ranging from neurotransmitter release to gene expression.

Nicotinic receptors are composed of five subunits (Fig 2.3). Each subunit is encoded by a single gene and denoted by a Greek letter and Arabic number.[14] All subunits expressed in the brain are either α or β (for example, α4, β2). Nicotinic receptors, in turn, can be differentiated into subtypes, depending on the combination of subunits that they contain. Thus, according to current nomenclature,[14] nAChRs that comprise only α4β2 subunits are termed α4β2 receptors; if additional types of subunits are suspected, a 'wild card' asterisk is added (hence α4β2* receptors). In the brain, α7* and α4β2* are particularly prevalent. It is not known how many nAChR subtypes are expressed in the mammalian brain, but pharmacological and molecular genetic studies have already identified about a dozen.[15] Nicotinic receptor subtypes differ greatly in their anatomical location, in their sensitivity to nicotine, and in their propensity to desensitise (lose responsiveness to nicotine) and resensitise (become responsive to nicotine again) when exposed to this drug.[15–17]

2.3 Brain nicotinic receptors and cigarette smoking

2.3.1 General issues

Nicotine is thought to play a critical role in tobacco dependence,[18] and since nicotinic ACh receptors represent the main molecular target for this drug, they are likely to be heavily involved in the mechanisms of nicotine dependence. The receptor subtype or subtypes relevant to dependence are likely to be those located in neuronal systems mediating reinforcement, of which the ventral tegmental area (VTA) appears to be particularly important (see Chapter 3). However, on current evidence, it is likely that nAChRs in other brain areas also contribute to nicotine dependence.

To be involved in the development of dependence, the relevant nAChR subtypes have to be sensitive to smoking-relevant concentrations of nicotine. It is important to note that nicotine and other agonists not only activate nicotinic receptors; these drugs can also turn off (desensitise) receptor function. In the desensitised state, receptors are temporarily unreactive and their ion channels are closed. It used to be thought that nicotinic receptors only assume a desensitised state when exposed to high, near-lethal concentrations of nicotine. However, this conclusion was drawn from work on peripheral nicotinic receptors. In the brain, nicotinic receptors are generally much more prone to desensitisation, and some can even be turned off by levels of nicotine experienced on a daily basis by smokers.

The common assumption is that smokers seek nicotine to cause nAChR stimulation rather than desensitisation, or possibly a combination of the two. If nicotine receptor activation is indeed an important component of why smokers smoke, it is therefore evident that if receptors desensitise on exposure to nicotine, they should regain their sensitivity rapidly enough to ensure that a significant proportion of the receptors can be activated at any one time.

Daytime venous plasma nicotine concentrations in smokers tend to peak at around 30 ng/ml or 0.2 µM,[1] and animal studies have shown that brain tissue sequesters nicotine, maintaining a 2.5–5-fold concentration gradient compared with levels in the blood.[19,20] Brain nicotine concentrations in habitual smokers are, therefore, likely to fluctuate between about 0.1 µM and 1 µM, though these estimates do not allow for sequestration of brain nicotine in lipid and other cellular compartments where it may be unavailable to act on receptors.

2.3.2 Differential sensitivity of nicotinic receptor subtypes to activation by nicotine

Acute sensitivity of nAChRs to nicotine has been assessed *in vitro* in recombinant nAChR subtypes expressed in cultured cells or in frog egg cells (oocytes) and in native (naturally occurring) nAChRs also isolated from mammalian tissue, and also *in vivo* in laboratory animals. Recombinant receptors represent a useful research tool, but they frequently fail to match their native counterparts in pharmacological and other respects.[21]

In isolated brain tissue, brief submicromolar application of nicotine to native nAChRs exerts numerous actions, including neuronal excitation and increased release of several neurotransmitters.[22,23] The use of isolated tissue permits greater control over the experimental conditions than is usually possible *in vivo*; consequently, one can be more confident that the observed effects of nicotine reflect direct actions of the drug on the neuronal population under study. The great majority of these *in vitro* studies have been conducted in rat or mouse tissue. Neuronal excitation has been investigated using electrophysiological recording techniques mainly in brain areas and neuronal populations that are known to express nAChRs abundantly. These include the hippocampus, midbrain dopamine neurons, thalamus, and medial habenular nucleus.[24–27] Electrophysiological approaches have also identified modulatory effects of nicotine on neurotransmitter release from defined neuronal projections in the brain, such as the thalamocortical and habenulo-interpeduncular pathways.[28–32] In most cases, a stimulatory effect of nicotine was observed, but inhibition has also been reported.[33] In most of these studies, it has been possible to identify the neurotransmitter in question, such as glutamate or gamma-aminobutyric acid (GABA).[28,29,32] Nicotine-evoked neurotransmitter release can also be studied from isolated synaptosomes (pinched-off nerve terminals) or tissue slices.[15,23] This approach is more versatile than electrophysiological recording in that it can detect the release of a wider range of neurotransmitters, notably dopamine, ACh, noradrenaline and serotonin, in addition to glutamate and GABA.[15,23]

The above effects of nicotine on neuronal excitability and neurotransmitter release are mediated by a variety of nAChR subtypes. For example, glutamate release is commonly regulated by nAChRs comprised of $\alpha 7$ subunits, whereas release of other neurotransmitters tends to be controlled by multiple receptor subtypes.[15] This diverse pharmacology is seen at nicotine concentrations that are relevant to smoking. Several nicotine-sensitive nAChR subtypes have been identified in isolated rodent tissue using pharmacological and genetic knockout approaches.[15,34–36] These include $\alpha 4 \beta 2^*$ nAChRs (see section 2.2 for explanation of nomenclature), which are widely and abundantly expressed in the brains of all mammalian species examined, including the human brain.[21] Other subtypes are distributed more discretely, such as $\alpha 6 \beta 3^*$ isoforms associated with dopamine releasing terminals.[34] In contrast, the $\alpha 7^*$ nAChR subtype, although abundant in humans and other mammals, is only weakly activated by nicotine when applied acutely in concentrations likely to occur during smoking.[21]

2.3.3 Modulation of nicotinic receptor function by desensitisation and resensitisation

Experiments in animals have shown that nicotinic receptors in the brain typically become at least partially unresponsive to nicotine (desensitised) when the drug is applied continuously for a few seconds or minutes. The speed at which this happens, referred to as desensitisation kinetics, helps to determine which potentially active nAChR subtypes predominate in cigarette smokers. Perhaps surprisingly, many receptors can be desensitised by concentrations of nicotine lower than those required for receptor activation.[17] Potentially, then, a proportion of brain nAChRs may be desensitised even before the first cigarette is smoked in the morning. *In vitro* studies have shown profound differences between nAChR subtypes in terms of the rapidity and extent to which they desensitise when acutely exposed to nicotine.[17] For example, α7 nAChRs desensitise in less than a second, but require high concentrations to do so, whereas α4β2 receptors can desensitise at much lower (nanomolar) concentrations of nicotine, albeit much more slowly.

Pharmacological studies have tended to focus on the acute activating and desensitising actions of nicotine. Nicotine AChR subtypes differ markedly in their propensity to desensitise to subchronic *in vitro* exposure to nicotine over hours or days. Based on a substantial literature, nicotine administered in concentrations of 0.1–0.3 μM has been estimated to reduce nAChR function by approximately 50% for α3β2-nAChR; by 10–50% for α3β4-nAChR; by 50–100% for α4β2-nAChR; by 20–50% for α4β4-nAChR; and by 50–100% for α7-nAChR.[37] Although not all of these receptor subtypes have yet been identified in the human brain, their existence seems likely, based on the anatomical distribution of messenger RNA (mRNA) that would encode them.[21,38] Accordingly, it has been suggested that, in smokers, α3*-nAChRs, being less apt to desensitise, may predominate functionally over α4* and α7*-nAChRs.[39] Such a conclusion must be viewed with caution, however, not least because it is based on the use of recombinant nAChRs, which quite frequently behave differently from their native counterparts.[37] In this context, one should also note that native α7 receptors have been identified in the embryonic rat brain that do not inactivate with subchronic nicotine exposure.[40]

While our knowledge of nAChR desensitisation kinetics is incomplete, even less is known about how rapidly different nAChR subtypes recover their function (resensitise) after brief nicotine exposure. Recombinant nAChRs have shown substantial recovery within minutes or hours of rapid nicotine removal in *in vitro* experiments, at least for the few receptor subtypes that have been tested thus far.[37,41] In human neuroblastoma cells, which possess native α3* and α7* subtypes, the full nicotinic response was regained within five minutes, even after 48 hours exposure to nicotine.[42] In isolated brain tissue, where multiple nAChR subtypes contribute to the response, some recovery of sensitivity has been observed within several minutes following brief application of nicotine,[43,44] but an additional persistent insensitivity lasting several hours has also been detected, even after acute administration.[45]

2.4 Insights into nAChR activation, desensitisation and resensitisation from *in vivo* studies

In vitro characterisation of nAChR functioning – activation, desensitisation and resensitisation – provides a starting point, but ultimately may tell us little about habitual smokers. One reason for this is that some potent nicotine actions observed *in vitro* have not been found to be significant when the drug is given acutely *in vivo*. This is illustrated by many dopaminergic neurons in the brain. These neurons express well-characterised nAChRs on their cell bodies and dendrites, which modulate cell firing.[36] In addition, dopaminergic neurons possess presynaptic receptors which promote transmitter release.[34] However, in laboratory rodents, systemic administration of nicotine stimulates dopamine release mainly or entirely by stimulating nAChRs in the dopamine cell body region.[46,47] Brain noradrenergic neurons provide a second example of this phenomenon.[48] *In vitro* studies are also subject to the caveat that they cannot readily reproduce the longer-term changes in nAChRs that result from prolonged *in vivo* exposure to nicotine. Such alterations are discussed in the next section.

2.5 Regulation of nicotinic receptor expression and function by chronic nicotine exposure

2.5.1 Differential regulation of nicotinic receptor subtypes

Chronic and continuous exposure to nicotine can markedly alter the expression and function of brain nAChRs *in vivo*. In terms of nAChR expression, post-mortem analysis has shown that cigarette smokers possess more nicotinic receptors than non-smokers and ex-smokers.[49–51] This high level of nAChR numbers was found in a number of brain regions, to somewhat differing degrees. Conceivably, individuals who ultimately become highly addicted smokers may possess more brain nicotinic receptors even before they take up cigarette smoking, or smoking may result in up-regulation of nicotinic receptors. The question of what is the main cause of the observed increase in nicotinic receptors in smokers could be examined best through longitudinal studies using brain imaging techniques. Animal studies strongly suggest that increased expression of nicotinic receptors stems at least in part from up-regulation following chronic nicotine administration.[37] Animal studies have also shown that nAChR subtypes differ in their propensity to proliferate in response to nicotine; for example, $\alpha4\beta2^*$-nAChRs up-regulate more readily than $\alpha7$-nAChRs and $\alpha3/\alpha6\beta2^*$-nAChRs.[37,52,53] Much less is known in this regard about the human brain, although, consistent with the animal literature, post-mortem analysis has revealed an increased density of $\alpha4\beta2^*$-nAChRs but not $\alpha7$-nAChRs in smokers.[50] The occurrence of agonist-induced receptor proliferation may appear paradoxical, but it is not always accompanied by increased receptor function.[37]

In the animal literature, there are numerous reports of either functional down- or up-regulation of receptors following chronic nicotine exposure.[37] Unfortunately, most studies of this sort have had to rely on post-mortem analysis of brain tissue, and some of the mixed results in this field may be attributable to differences in the way that brain tissue is prepared. Nevertheless, when tissue preparation is standardised, it becomes apparent that nicotinic receptors located on different neuronal populations can regulate in opposite directions. For example, in one study, chronic nicotine administration led to an increase in nicotine-evoked release of dopamine and 5-HT from rat striatal slices, whereas ACh release was reduced in the same tissue.[54] In another more recent study, chronic nicotine administration greatly attenuated the ability of an acute nicotine challenge to promote *ex vivo* hippocampal ACh release, whereas nicotine-evoked noradrenaline release was increased.[55]

These diverse findings presumably reflect anatomical differences in receptor subtype expression. This is suggested by studies in cultured cells, where extended nicotine exposure alters nAChR expression and function in a highly subtype-dependent manner.[37] Hence, it seems likely that chronic nicotine exposure in smokers promotes the function of certain nAChR subtypes at the expense of others, but the literature from animal studies gives little indication of which subtypes may be favoured in this way. For example, while radioligand binding studies in rats have reported a selective proliferation of $\alpha 4 \beta 2^*$ or $\alpha 6^*$ receptors,[56,57] it is not yet established whether increased receptor abundance translates into increased receptor function in these cases.

With *in vivo* nicotine exposure, more persistent forms of nAChR desensitisation ('receptor inactivation') are likely to come to the fore. Evidence for this comes from rodent studies where as few as one or two systemic injections of nicotine can induce nAChR-mediated tolerance for a few or even many hours, depending on the study.[58–60] However, multiple injections of nicotine given over several days have been found to produce a much more prolonged nAChR inactivation, such that the prolactin release response to nicotine did not recover for more than a week after the last injection.[58]

It is still largely a matter of speculation which of the many nAChR populations in the brain remain active during daytime smoking. Recently, however, it has been claimed that most $\alpha 4 \beta 2$ receptors are desensitised by cigarette smoking.[61] These authors used a positron emission tomography (PET) imaging radiotracer to estimate the degree of occupancy of $\alpha 4 \beta 2$-nAChRs in human smokers. Given that radio-labelled nicotine binds with high (nanomolar) affinity to $\alpha 4 \beta 2$ receptors *in vitro*, it was hypothesised that these receptors would be almost completely occupied by cigarette smoking. Subjects were screened to confirm that they had abstained from cigarettes for at least 48 hours when tested. Smoking exerted a remarkably profound effect on radiotracer binding, with effects apparent even after a single cigarette puff, and it was estimated that daytime smoking results in near-complete receptor occupancy (96% or more). It was further suggested that, across the day, most α4β2 receptors shift to a high-affinity desensitised state, leaving very few that can still be activated by nicotine. However, this conclusion is tempered by the caveat that nAChR radiotracers are not capable of detecting receptors remaining in a lower-affinity state which would still be possible to activate by nicotine.

2.5.2 Effect of chronic exposure on acute *in vivo* effects of nicotine in animals

Nicotine exerts numerous acute neuropharmacological and behavioural effects in drug-naive animals, but few studies have examined whether these effects can be obtained during chronic exposure to nicotine. This is surprising, since it is widely acknowledged that most cigarette smokers are continuously exposed to pharmacological concentrations of nicotine, even during the night and before the first cigarette of the day.[1] For various reasons, it has also proved difficult to emulate the pronounced diurnal pattern of nicotine exposure experienced by smokers in laboratory animals. Most investigators have employed subcutaneous osmotic mini-pumps that deliver nicotine at a constant rate over several days up to a few weeks. Against a background of continuous nicotine exposure, acute nicotine administration lost its ability to increase detectably dopamine overflow in the nucleus accumbens, measured by the microdialysis sampling method in rats.[62,63] This finding may be highly significant, given the importance currently attributed to the mesolimbic dopamine system in mediating the reinforcing effects of nicotine (see Chapter 3). Importantly, the observed insensitivity to nicotine was associated with plasma concentrations mimicking daytime smoking levels. The loss of nicotine-induced mesolimbic dopamine overflow persisted for at least one or two days after withdrawal of chronic nicotine in rats.[64]

By no means are all of nicotine's acute effects lost during chronic drug exposure, however. For example, in the above-mentioned microdialysis experiments, chronic drug infusion that was sufficient to abolish the mesolimbic dopamine response had little if any effect on dopamine overflow in the neighbouring nigrostriatal dopamine pathway.[62] Clearly, then, at least some acute actions of nicotine can persist despite continuous nicotine dosing. This fact has also been demonstrated using immunohistochemical mapping of fos, a protein marker for neuronal activation.[65,66]

In drug-naive animals, a single systemic nicotine injection increased fos expression in several brain areas, as expected from previous studies. This acute response was affected variably by continuous chronic nicotine delivered by osmotic mini-pump; in some brain areas (the nucleus accumbens and parts of the hypothalamus, for example) it was reduced or abolished, whereas in another area (the central nucleus of the amygdala) it was unaffected.[65,66] It is not known why tolerance to nicotine should develop in some neuronal populations but not in others, but differences in nicotinic receptor subtype expression almost certainly contribute to these differences.

The persistence of functional brain nAChRs in fos expression studies is remarkable, considering that chronic plasma nicotine levels in these studies were well in excess of those found in typical smokers. However, at least two other types of study provide corroborating evidence. In the first, administration of a nAChR antagonist to rodents receiving chronic nicotine readily precipitates withdrawal signs, some of which are mediated by brain nAChRs;[67] such a result would not be expected if all nAChRs were already inactivated. Second, nicotine-induced stimulation of brain nAChRs provides a signal that rats can learn to recognise in order to make the correct response to obtain a food reward. This acute effect of nicotine persists undiminished when superimposed on a background of chronic continuous nicotine.[68]

2.6 Acute actions of nicotine maintained in habitual cigarette smokers

The evidence from animal studies reviewed above suggests, therefore, that, in the continued presence of nicotine, some brain nAChRs lose their ability to function while others remain sensitive to nicotine. But is this also the case in human cigarette smokers, and what happens to the key receptors which may modulate neurobiological processes underlying addiction to nicotine? These are harder questions to answer and few published studies have addressed this issue. It is well established that cigarette smokers become partially tolerant to a number of nicotine's subjective and behavioural effects, most notably nausea and other aversive effects.[69] As in animal studies, short-term tolerance can be demonstrated within minutes or hours of acute drug administration, and more prolonged drug exposure results in chronic tolerance. It is likely that at least some of the tolerance observed is mediated by desensitisation of nAChRs in the brain.

The animal evidence reviewed above suggests that profound tolerance to the actions of nicotine develops in some brain areas but not in others. In human subjects, it is possible to visualise the anatomical pattern of nicotinic activation using functional magnetic resonance imaging (fMRI). Such a study has revealed widespread effects of intravenous nicotine in cigarette smokers.[70] However, the nicotine doses used in that study were extremely high, equivalent to the nicotine yield from one to two cigarettes infused in a single minute.[1] Hence, it is possible that nicotine was able to stimulate nAChR subtypes that would not normally be activated during smoking.

Human brain imaging has also been used to estimate the proportion of α4β2 brain nicotinic receptors that are occupied by circulating levels of nicotine.[61] The investigators concluded that, across the day, cigarette smoking causes most α4β2 receptors to become desensitised, leaving very few that can still be activated by nicotine. As mentioned above, this conclusion should perhaps be viewed with caution, but, if true, these findings are potentially important given the high prevalence of this receptor subtype in the brain and in dopaminergic pathways in particular.

Perhaps surprisingly, administration of the nAChR blocker mecamylamine does not produce significant signs of nicotine withdrawal in human smokers, even when they are actively smoking.[71,72] Since mecamylamine is centrally active, the absence of withdrawal effects might suggest that few nAChRs are still functional in these individuals. However, mecamylamine can precipitate mild withdrawal in human subjects wearing a nicotine patch instead of smoking.[73] Administration of mecamylamine also increases acutely cigarette smoking in a consistent manner, as if smokers are attempting to overcome the effects of nAChR blockade.[71,73,74] Unfortunately, mecamylamine antagonises both central and peripheral nAChRs, both of which contribute to the short-term regulation of smoking behaviour.[75] Hence, human studies with mecamylamine do not unequivocally demonstrate the presence of functional nAChRs in the brain of active smokers.

2.7 Effects of nicotine on release of brain dopamine in smokers

It is widely thought that cigarette smokers seek nicotine because it releases dopamine in the mesolimbic system of the brain. Most of the evidence derives from laboratory studies in rodents. Thus, systemic injection of nicotine reliably increases mesolimbic dopamine release (see Chapter 3), and the same phenomenon has been observed in animals actively working to obtain the drug.[76] In addition, laboratory animals self-administer nicotine less if dopaminergic transmission is disrupted surgically or with drugs.[77,78]

This sort of evidence may appear compelling, but it is important to consider the limitations of such studies, apart from the obvious possibility of species differences. First, animals were typically drug-free when given nicotine, so brain nAChRs were minimally desensitised. Second, most of these studies were of only a few days' duration, reducing the possibility of compensatory receptor up- or down-regulation of nAChR density. Third, rats self-administering the drug were offered only limited daily access, insufficient to produce dependence.[79] Fourth, self-administered nicotine was delivered in multiple doses, each equivalent to the yield from one whole cigarette, potentially producing actions in the brain that would not be encountered by human smokers. Finally, intravenous nicotine was injected almost instantaneously, whereas nicotine from cigarette smoke is released into the bloodstream over the course of a minute or so.[2] This difference could be critical because recent evidence suggests that such differences in infusion speed can have a great effect on the impact that nicotine has on the brain in animal models.[80]

What then is the evidence that cigarette smoking increases mesolimbic dopamine release (in human subjects)? This question has been addressed principally with PET imaging, using radiotracers that bind to dopamine receptors; the extent of dopamine release is inferred indirectly from a decline in radiotracer labelling. Results to date have been somewhat mixed. A first study, targeting the D1 subtype of dopamine receptor, provided no evidence for smoking-induced dopamine release.[81] However, in this study, the control subjects were expecting a cigarette immediately after the scan, so it is possible that expectation-induced dopamine release obscured any effect of active smoking.

Subsequent studies have focused on the D2 dopamine receptor subtype. In one report,[82] dopamine release did not consistently increase during cigarette smoking in subjects who were at least 12 hours abstinent. However, the degree of pleasure experienced during smoking was correlated with dopamine release. Unexpectedly, this correlation occurred not in the mesolimbic system, but instead in a dopamine terminal area (caudate-putamen) more usually associated with motor control than with emotion. In contrast, another group observed an apparent increase in mesolimbic dopamine release in subjects smoking after only three hours of deprivation.[83] This effect was surprisingly large, more akin to that seen previously with amphetamine and cocaine. However, a much smaller effect of smoking has now been reported in a new group of subjects.[84]

Taken together, these PET imaging studies suggest that in humans cigarette smoke can increase dopamine release, even if the location and extent of this effect appears somewhat unpredictable. All subjects in the studies were at least three hours abstinent from smoking, and such an interval could conceivably allow many desensitised nAChRs to regain their function prior to the test session. Perhaps, then, the test cigarette resembled most closely that of the first cigarette in a normal smoker's day. What remains entirely unknown is whether subsequent cigarettes are also capable of releasing brain dopamine. Instead, smoking-associated cues may well play a key role in maintaining cigarette smoking behaviour (see Chapter 3).

A further limitation of brain imaging studies is that they cannot tell us directly how much of an impact any dopamine release would have on smoking behaviour or subjective responses to smoking. One way to address this question is to attenuate dopaminergic transmission experimentally by giving a dopamine receptor blocker. This approach has yielded mixed results, with both increases and decreases in smoking behaviour reported in different studies.[85–87] Here, interpretation is further complicated by the possibility of non-selective drug effects unrelated to dopamine receptor blockade. Dopaminergic transmission can also be reduced temporarily in human subjects by administering an experimental meal lacking the chemical precursors for this neurotransmitter. This dietary manipulation has previously been found to decrease the stimulant effects of amphetamine and cocaine – drugs which normally increase dopaminergic transmission. However, the dopamine-depleting diet failed to change smoking behaviour, the urge to smoke or various subjective effects of smoking.[88]

2.8 Contribution of nicotinic receptor subtypes to the reinforcing effects of nicotine in animal models

Nicotinic receptors are distributed widely in the brain, and their natural function is to mediate the effects of the neurotransmitter ACh. Since nAChRs may be physiologically important, smoking cessation drugs that target multiple nAChRs indiscriminately are likely to have unwanted effects. With a view to developing more selective drugs, much effort has been expended trying to identify subtypes of nAChR that contribute to the reinforcing effects of nicotine in animal models.[89] Behavioural tests have focused largely on conditioned place preference (CPP) and intravenous nicotine self-administration (IVSA), with rats and mice being the animals of choice. The CPP procedure relies on the tendency of animals to prefer an environment where they previously received a rewarding stimulus. This type of conditioning is usually carried out using a two-compartment box: the animal is repeatedly confined to one side after drug injection and to the other side after a control injection of saline. On the test day, both sides are accessible and the time spent in each is recorded. The IVSA procedure, in contrast, makes use of a Skinner box equipped with levers, and here the animal must work in order to obtain the drug. Attention has also been paid to the mesolimbic dopamine system, in the expectation that this neuronal population mediates the drug's reinforcing effects. Since nicotinic drugs tend to have limited selectivity for a given receptor subtype, much of this research has relied on the use of genetically modified (especially 'knockout') mice that fail to express a particular nAChR subunit.[90]

Several conclusions can be drawn from these studies. First, CPP associated with systemic nicotine administration requires the expression of α4 and β2 but not α7 subunits.[91,92] Second, dopamine release evoked by acute systemic injection of nicotine also requires the presence of α4 and β2 nAChR subunits,[93,94] probably located in the dopamine cell body area.[34,36,95] Third, nicotine IVSA relies to an important extent on nAChRs containing the β2 subunit[93,96] and possibly depends on α7 subunits as well, although here results are mixed.[97,98] The participation of α4-containing receptors in nicotine IVSA has not been established but seems likely, since this behaviour is dependent on mesolimbic dopamine transmission.[77] Finally, the possible participation of other nAChR subunits has not yet been reported.

Signs of nicotine withdrawal have also been studied in mice lacking specific nAChR subunits. In one study, mice were chronically treated with oral nicotine so as to achieve plasma levels in the low smoking range.[99] Spontaneous withdrawal subsequently elicited several behavioural signs, some being dependent on the expression of α7 subunits. Other studies have examined withdrawal precipitated by acute administration of the antagonist mecamylamine during chronic nicotine administration.[100,101] Here, β4 but not α2 receptors appeared critical, but circulating levels of nicotine were not measured to see whether they were in the smoking range.

Thus, it is still not clear which nAChR subtype or subtypes contribute to tobacco dependence, but animal models suggest that nicotinic receptors containing α4 and β2 subunits are the most likely candidates. This conclusion appears consistent with the recently reported success of the α4β2-selective agonist varenicline in smoking cessation trials, although it should be borne in mind that varenicline has additional pharmacological actions.[102] Finally, it is important to emphasise that existing animal testing paradigms model, at best, only certain aspects of cigarette smoking.

2.9 Conclusions

▸ Nicotine targets receptors whose natural function is to interact with the neurotransmitter acetylcholine (ACh).

▸ By activating these nicotinic receptors, nicotine increases the firing rate of neurons and increases the release of various neurotransmitters.

▶ The effect of nicotine on different nicotinic receptors is dose dependent and is also modified by sustained exposure, which causes some receptors to become desensitised.

▶ Long-term exposure to nicotine also causes an increase in the number of nicotinic receptors in the brain, but it is not clear how many of these receptors are functional.

▶ In animal studies, the acute reinforcing effects of nicotine appear to be dependent on dopamine release in the brain. However, nicotine-induced dopamine release is curtailed markedly or even disappears when animals are chronically exposed to the drug.

▶ Although cigarette smoking in humans promotes dopamine release, the contribution of this effect to sustained smoking behaviour in humans is still not fully understood (see Chapter 3).

▶ It is also not clear which nicotinic receptor subtypes are responsible for reinforcing effects of this drug in humans.

References

1   Hukkanen J, Jacob P, Benowitz NL. Metabolism and disposition kinetics of nicotine. *Pharmacol Rev* 2005;57:79–115.

2   Rose JE, Behm FM, Westman EC, Coleman RE. Arterial nicotine kinetics during cigarette smoking and intravenous nicotine administration: implications for addiction. *Drug Alcohol Depend* 1999;56:99–107.

3   Bradbury MW, Patlak CS, Oldendorf WH. Analysis of brain uptake and loss or radiotracers after intracarotid injection. *Am J Physiol* 1975;229:1110–5.

4   Nybäck H, Halldin C, Ahlin A *et al.* PET studies of the uptake of (S)- and (R)-[^{11}C]nicotine in the human brain: difficulties in visualizing specific receptor binding *in vivo. Psychopharmacology* 1994;115:31–6.

5   Benowitz NL, Kuyt F, Jacob P. Circadian blood nicotine concentrations during cigarette smoking. *Clin Pharmacol Ther* 1982;32:758–64.

6   Benowitz NL, Kuyt F, Jacob P, Jones RT, Osman AL. Cotinine disposition and effects. *Clin Pharmacol Ther* 1983;34:604–11.

7   Ghosheh OA, Dwoskin LP, Miller DK, Crooks PA. Accumulation of nicotine and its metabolites in rat brain after intermittent or continuous peripheral administration of [2'- (14)C]nicotine. *Drug Metab Dispos* 2001;29:645–51.

8   Crooks PA, Dwoskin LP. Contribution of CNS nicotine metabolites to the neuropharmacological effects of nicotine and tobacco smoking. *Biochem Pharmacol* 1997;54: 743–53.

9   Teng LH, Crooks PA, Buxton ST, Dwoskin LP. Nicotinic-receptor mediation of S(-)nornicotine-evoked [^{3}H]overflow from rat striatal slices preloaded with [^{3}H]dopamine. *J Pharmacol Exp Ther* 1997;283:778–87.

10 Dwoskin LP, Teng LH, Buxton ST, Crooks PA. (S)-(-)-cotinine, the major brain metabolite of nicotine, stimulates nicotinic receptors to evoke [^{3}H]dopamine release from rat striatal slices in a calcium-dependent manner. *J Pharmacol Exp Ther* 1999;288:905–11.

11   Bardo MT, Green TA, Crooks PA, Dwoskin LP. Nornicotine is self-administered intravenously by rats. *Psychopharmacology* 1999;146:290–6.

12   Clarke PBS, Schwartz RD, Paul SM, Pert CB, Pert A. Nicotinic binding in rat brain: auto-radiographic comparison of ^{3}H-acetylcholine, ^{3}H-nicotine, and ^{125}I-alpha-bungarotoxin. *J Neurosci* 1985;5:1307–15.

13   Wada E, Wada K, Boulter J *et al*. Distribution of alpha 2, alpha 3, alpha 4, and beta 2 neuronal nicotinic receptor subunit mRNAs in the central nervous system: a hybridization histochemical study in the rat. *J Comp Neurol* 1989;284:314–35.

14   Lukas RJ, Changeux JP, Le Novère N *et al*. International Union of Pharmacology. XX. Current status of the nomenclature for nicotinic acetylcholine receptors and their subunits. *Pharmacol Rev* 1999;51:397–401.

15   Gotti C, Zoli M, Clementi F. Brain nicotinic acetylcholine receptors: native subtypes and their relevance. *Trends Pharmacol Sci* 2006;27:482–91.

16   Royal College of Physicians. *Nicotine addiction in Britain*. Report of the Tobacco Advisory Group of the Royal College of Physicians. London: RCP, 2000.

17   Giniatullin R, Nistri A, Yakel JL. Desensitization of nicotinic ACh receptors: shaping cholinergic signaling. *Trends Neurosci* 2005;28:371–8.

18   Stolerman IP, Jarvis MJ. The scientific case that nicotine is addictive. *Psychopharmacology* 1995;117:2–10.

19   Rowell PP, Li M. Dose-response relationship for nicotine-induced up-regulation of rat brain nicotinic receptors. *J Neurochem* 1997;68:1982–9.

20   Decker MW, Brioni JD, Sullivan JP *et al*. (S)-3-methyl-5-(1-methyl-2-pyrrolidinyl)isoxazole (ABT 418): a novel cholinergic ligand with cognition-enhancing and anxiolytic activities: II. *In vivo* characterization. *J Pharmacol Exp Ther* 1994;270:319–28.

21   Gotti C, Fornasari D, Clementi F. Human neuronal nicotinic receptors. *Prog Neurobiol* 1997;53:199–237.

22   Jones S, Sudweeks S, Yakel JL. Nicotinic receptors in the brain: correlating physiology with function. *Trends Neurosci* 1999;22:555–61.

23   Wonnacott S. Presynaptic nicotinic ACh receptors. *Trends Neurosci* 1997;20:92–8.

24   Pidoplichko VI, DeBiasi M, Williams JT, Dani JA. Nicotine activates and desensitizes midbrain dopamine neurons. *Nature* 1997;390:401–4.

25   Albuquerque EX, Alkondon M, Pereira EFR *et al*. Properties of neuronal nicotinic acetylcholine receptors: Pharmacological characterization and modulation of synaptic function. *J Pharmacol Exp Ther* 1997;280:1117–36.

26   McCormick DA, Prince DA. Acetylcholine causes rapid nicotinic excitation in the medial habenular nucleus of guinea pig, *in vitro*. *J Neurosci* 1987;7:742–52.

27   McCormick DA, Prince DA. Actions of acetylcholine in the guinea-pig and cat medial and lateral geniculate nuclei, *in vitro*. *J Physiol (Lond)* 1987;392:147–65.

28   Gray R, Rajan AS, Radcliffe KA *et al*. Hippocampal synaptic transmission enhanced by low concentrations of nicotine. *Nature* 1996;383:713–6.

29   Mansvelder HD, Keath JR, McGehee DS. Synaptic mechanisms underlie nicotine-induced excitability of brain reward areas. *Neuron* 2002;33:905–19.

30   Gioanni Y, Rougeot C, Clarke PBS *et al*. Nicotinic receptors in the rat prefrontal cortex: increase in glutamate release and facilitation of mediodorsal thalamo-cortical transmission. *Eur J Neurosci* 1999;11:18–30.

31   Guo JZ, Tredway TL, Chiappinelli VA. Glutamate and GABA release are enhanced by different subtypes of presynaptic nicotinic receptors in the lateral geniculate nucleus. *J Neurosci* 1998;18:1963–9.

32   Metherate R. Nicotinic acetylcholine receptors in sensory cortex. *Learn Mem* 2004;11:50–9.

33   Mulle C, Vidal C, Benoit P, Changeux JP. Existence of different subtypes of nicotinic acetylcholine receptors in the rat habenulo-interpeduncular system. *J Neurosci* 1991;11: 2588–97.

34   Champtiaux N, Gotti C, Cordero-Erausquin M *et al.* Subunit composition of functional nicotinic receptors in dopaminergic neurons investigated with knock-out mice. *J Neurosci* 2003;23:7820–9.

35   Cui C, Booker TK, Allen RS *et al.* The beta3 nicotinic receptor subunit: a component of alpha-conotoxin MII-binding nicotinic acetylcholine receptors that modulate dopamine release and related behaviors. *J Neurosci* 2003;23:11045–53.

36   Mameli-Engvall M, Evrard A, Pons S *et al.* Hierarchical control of dopamine neuron-firing patterns by nicotinic receptors. *Neuron* 2006;50:911–21.

37   Gentry CL, Lukas RJ. Regulation of nicotinic acetylcholine receptor numbers and function by chronic nicotine exposure. *Curr Drug Targets CNS Neurol Disord* 2002;1: 359–85.

38   Court JA, Martin-Ruiz C, Graham A, Perry E. Nicotinic receptors in human brain: topography and pathology. *J Chem Neuroanat* 2000;20:281–98.

39   Olale F, Gerzanich V, Kuryatov A *et al.* Chronic nicotine exposure differentially affects the function of human alpha3, alpha4, and alpha7 neuronal nicotinic receptor subtypes. *J Pharmacol Exp Ther* 1997;283:675–83.

40   Kawai H, Berg DK. Nicotinic acetylcholine receptors containing alpha7 subunits on rat cortical neurons do not undergo long-lasting inactivation even when up-regulated by chronic nicotine exposure. *J Neurochem* 2001;78:1367–78.

41   Quick MW, Lester RA. Desensitization of neuronal nicotinic receptors. *J Neurobiol* 2002;53:457–78.

42   Sokolova E, Matteoni C, Nistri A. Desensitization of neuronal nicotinic receptors of human neuroblastoma SH-SY5Y cells during short or long exposure to nicotine. *Br J Pharmacol* 2005;146:1087–95.

43   Marks MJ, Grady SR, Yang J-M *et al.* Desensitization of nicotine-stimulated ^{86}Rb$^+$ efflux from mouse brain synaptosomes. *J Neurochem* 1994;63:2125–35.

44   Grady SR, Marks MJ, Collins AC. Desensitization of nicotine-stimulated [^{3}H]dopamine release from mouse striatal synaptosomes. *J Neurochem* 1994;62:1390–8.

45   Rowell PP, Duggan DS. Long-lasting inactivation of nicotinic receptor function *in vitro* by treatment with high concentrations of nicotine. *Neuropharmacology* 1998;37:103–11.

46   Benwell MEM, Balfour DJK, Lucchi HM. Influence of tetrodotoxin and calcium on changes in extracellular dopamine levels evoked by systemic nicotine. *Psychopharmacology* 1993;112:467–74.

47   Nisell M, Nomikos GG, Svensson TH. Systemic nicotine-induced dopamine release in the rat nucleus accumbens is regulated by nicotinic receptors in the ventral tegmental area. *Synapse* 1994;16:36–44.

48   Matta SG, McCoy JG, Foster CA, Sharp BM. Nicotinic agonists administered into the fourth ventricle stimulate norepinephrine secretion in the hypothalamic paraventricular nucleus: An *in vivo* microdialysis study. *Neuroendocrinology* 1995;61:383–92.

49   Benwell ME, Balfour DJ, Anderson JM. Evidence that tobacco smoking increases the density of (-)- [^{3}H]nicotine binding sites in human brain. *J Neurochem* 1988;50:1243–7.

50   Court JA, Lloyd S, Thomas N *et al.* Dopamine and nicotinic receptor binding and the levels of dopamine and homovanillic acid in human brain related to tobacco use. *Neuroscience* 1998;87:63–78.

51   Breese CR, Marks MJ, Logel J *et al.* Effect of smoking history on [3H]nicotine binding in human postmortem brain. *J Pharmacol Exp Ther* 1997;282:7–13.

52   El-Bizri H, Clarke PBS. Regulation of nicotinic receptors in rat brain following quasi-irreversible nicotinic blockade by chlorisondamine and chronic treatment with nicotine. *Br J Pharmacol* 1994;113:917–25.

53   McCallum SE, Parameswaran N, Bordia T *et al.* Differential regulation of mesolimbic alpha 3/alpha 6 beta 2 and alpha 4 beta 2 nicotinic acetylcholine receptor sites and function after long-term oral nicotine to monkeys. *J Pharmacol Exp Ther* 2006;318:381–8.

54   Yu ZJ, Wecker L. Chronic nicotine administration differentially affects neurotransmitter release from rat striatal slices. *J Neurochem* 1994;63:186–94.

55   Grilli M, Parodi M, Raiteri M, Marchi M. Chronic nicotine differentially affects the function of nicotinic receptor subtypes regulating neurotransmitter release. *J Neurochem* 2005;93:1353–60.

56   Nguyen HN, Rasmussen BA, Perry DC. Subtype-selective up-regulation by chronic nicotine of high-affinity nicotinic receptors in rat brain demonstrated by receptor autoradiography. *J Pharmacol Exp Ther* 2003;307:1090–7.

57   Parker SL, Fu Y, McAllen K *et al.* Up-regulation of brain nicotinic acetylcholine receptors in the rat during long-term self-administration of nicotine: disproportionate increase of the alpha6 subunit. *Mol Pharmacol* 2004;65:611–22.

58   Hulihan-Giblin BA, Lumpkin MD, Kellar KJ. Effects of chronic administration of nicotine on prolactin release in the rat: inactivation of prolactin response by repeated injections of nicotine. *J Pharmacol Exp Ther* 1990;252:21–5.

59   Fu YT, Matta SG, Valentine JD, Sharp BM. Desensitization and resensitization of norepinephrine release in the rat hippocampus with repeated nicotine administration. *Neurosci Lett* 1998;241:147–50.

60   Vann RE, James JR, Rosecrans JA, Robinson SE. Nicotinic receptor inactivation after acute and repeated *in vivo* nicotine exposures in rats. *Brain Res* 2006;1086:98–103.

61   Brody AL, Mandelkern MA, London ED *et al.* Cigarette smoking saturates brain alpha 4 beta 2 nicotinic acetylcholine receptors. *Arch Gen Psychiatry* 2006;63:907–15.

62   Benwell MEM, Balfour DJK. Regional variation in the effects of nicotine on catecholamine overflow in rat brain. *Eur J Pharmacol* 1997;325:13–20.

63   Benwell MEM, Balfour DJK, Birrell CE. Desensitization of the nicotine-induced mesolimbic dopamine responses during constant infusion with nicotine. *Br J Pharmacol* 1995;114:454–60.

64   Rahman S, Zhang J, Engleman EA, Corrigall WA. Neuroadaptive changes in the mesoaccumbens dopamine system after chronic nicotine self-administration: a microdialysis study. *Neuroscience* 2004;129:415–24.

65   Salminen O, Seppa T, Gaddnas H, Ahtee L. Effect of acute nicotine on Fos protein expression in rat brain during chronic nicotine and its withdrawal. *Pharmacol Biochem Behav* 2000;66:87–93.

66   Salminen O, Seppä T, Gäddnäs H, Ahtee L. The effects of acute nicotine on the metabolism of dopamine and the expression of Fos protein in striatal and limbic brain areas of rats during chronic nicotine infusion and its withdrawal. *J Neurosci* 1999;19: 8145–51.

67   Kenny PJ, Markou A. Neurobiology of the nicotine withdrawal syndrome. *Pharmacol Biochem Behav* 2001;70:531–49.

68   Shoaib M, Thorndike E, Schindler CW, Goldberg SR. Discriminative stimulus effects of nicotine and chronic tolerance. *Pharmacol Biochem Behav* 1997;56:167–73.

69   Perkins KA. Chronic tolerance to nicotine in humans and its relationship to tobacco dependence. *Nicotine Tob Res* 2002;4:405–22.

70   Stein EA, Pankiewicz J, Harsch HH *et al*. Nicotine-induced limbic cortical activation in the human brain: A functional MRI study. *Am J Psychiatry* 1998;155:1009–15.

71   Nemeth Coslett R, Henningfield JE, O'Keeffe MK, Griffiths RR. Effects of mecamylamine on human cigarette smoking and subjective ratings. *Psychopharmacology* 1986;88:420–5.

72   Rose JE, Behm FM, Westman EC. Nicotine-mecamylamine treatment for smoking cessation: the role of pre-cessation therapy. *Exp Clin Psychopharmacol* 1998;6:331–43.

73   Rose JE, Behm FM, Westman EC. Acute effects of nicotine and mecamylamine on tobacco withdrawal symptoms, cigarette reward and ad lib smoking. *Pharmacol Biochem Behav* 2001;68:187–97.

74   Stolerman IP, Goldfarb T, Fink R, Jarvik ME. Influencing cigarette smoking with nicotine antagonists. *Psychopharmacologia* 1973;28:247–59.

75   Rose JE. Nicotine and nonnicotine factors in cigarette addiction. *Psychopharmacology* 2006;184:274–85.

76   Lecca D, Cacciapaglia F, Valentini V *et al*. Preferential increase of extracellular dopamine in the rat nucleus accumbens shell as compared to that in the core during acquisition and maintenance of intravenous nicotine self-administration. *Psychopharmacology* 2006;184:435–46.

77   Corrigall WA, Franklin KBJ, Coen KM, Clarke PBS. The mesolimbic dopaminergic system is implicated in the reinforcing effects of nicotine. *Psychopharmacology* 1992;107:285–9.

78   Corrigall WA, Coen KM. Selective dopamine antagonists reduce nicotine self-administration. *Psychopharmacology* 1991;104:171–6.

79   Paterson NE, Markou A. Prolonged nicotine dependence associated with extended access to nicotine self-administration in rats. *Psychopharmacology* 2004;173:64–72.

80   Samaha AN, Yau WY, Yang P, Robinson TE. Rapid delivery of nicotine promotes behavioral sensitization and alters its neurobiological impact. *Biol Psychiatry* 2005;57: 351–60.

81   Dagher A, Bleicher C, Aston JA *et al*. Reduced dopamine D1 receptor binding in the ventral striatum of cigarette smokers. *Synapse* 2001;42:48–53.

82   Barrett SP, Boileau I, Okker J, Pihl RO, Dagher A. The hedonic response to cigarette smoking is proportional to dopamine release in the human striatum as measured by positron emission tomography and [11C]raclopride. *Synapse* 2004;54:65–71.

83   Brody AL, Olmstead RE, London ED *et al*. Smoking-induced ventral striatum dopamine release. *Am J Psychiatry* 2004;161:1211–8.

84   Brody AL, Mandelkern MA, Olmstead RE *et al*. Gene Variants of Brain Dopamine Pathways and Smoking-Induced Dopamine Release in the Ventral Caudate/Nucleus Accumbens. *Arch Gen Psychiatry* 2006;63:808–16.

85   Brauer LH, Cramblett MJ, Paxton DA, Rose JE. Haloperidol reduces smoking of both nicotine-containing and denicotinized cigarettes. *Psychopharmacology* 2001;159:31–7.

86   Caskey NH, Jarvik ME, Wirshing WC. The effects of dopaminergic D2 stimulation and blockade on smoking behavior. *Exp Clin Psychopharmacol* 1999;7:72–8.

87   Dawe S, Gerada C, Russell MAH, Gray JA. Nicotine intake in smokers increases following a single dose of haloperidol. *Psychopharmacology* 1995;117:110–5.

88   Casey KF, Benkelfat C, Young SN, Leyton M. Lack of effect of acute dopamine precursor depletion in nicotine-dependent smokers. *Eur Neuropsychopharmacol* 2006;16:512–20.

89   Wonnacott S, Sidhpura N, Balfour DJ. Nicotine: from molecular mechanisms to behaviour. *Curr Opin Pharmacol* 2005;5:53–9.

90 Drago J, McColl CD, Horne MK, Finkelstein DI, Ross SA. Neuronal nicotinic receptors: insights gained from gene knockout and knockin mutant mice. *Cell Mol Life Sci* 2003;60: 1267–80.

91 Tapper AR, McKinney SL, Nashmi R *et al.* Nicotine activation of alpha4* receptors: sufficient for reward, tolerance, and sensitization. *Science* 2004;306:1029–32.

92 Walters CL, Brown S, Changeux JP *et al.* The beta2 but not alpha7 subunit of the nicotinic acetylcholine receptor is required for nicotine-conditioned place preference in mice. *Psychopharmacology* 2006;184:339–44.

93 Picciotto MR, Zoli M, Rimondini R *et al.* Acetylcholine receptors containing the b2 subunit are involved in the reinforcing properties of nicotine. *Nature* 1998;391:173–7.

94 Marubio LM, Gardier AM, Durier S *et al.* Effects of nicotine in the dopaminergic system of mice lacking the alpha4 subunit of neuronal nicotinic acetylcholine receptors. *Eur J Neurosci* 2003;17:1329–37.

95 Maskos U, Molles BE, Pons S *et al.* Nicotine reinforcement and cognition restored by targeted expression of nicotinic receptors. *Nature* 2005;436:103–7.

96 Epping-Jordan MP, Picciotto MR, Changeux JP, Pich EM. Assessment of nicotinic acetylcholine receptor subunit contributions to nicotine self-administration in mutant mice. *Psychopharmacology* 1999;147:25–6.

97 Markou A, Paterson NE. The nicotinic antagonist methyllycaconitine has differential effects on nicotine self-administration and nicotine withdrawal in the rat. *Nicotine Tob Res* 2001;3:361–73.

98 Grottick AJ, Trube G, Corrigall WA *et al.* Evidence that nicotinic alpha(7) receptors are not involved in the hyperlocomotor and rewarding effects of nicotine. *J Pharmacol Exp Ther* 2000;294:1112–9.

99 Grabus SD, Martin BR, Imad DM. Nicotine physical dependence in the mouse: involvement of the alpha7 nicotinic receptor subtype. *Eur J Pharmacol* 2005;515:90–3.

100 Salas R, Pieri F, De Biasi M. Decreased signs of nicotine withdrawal in mice null for the beta4 nicotinic acetylcholine receptor subunit. *J Neurosci* 2004;24:10035–9.

101 Besson M, David V, Suarez S *et al.* Genetic dissociation of two behaviors associated with nicotine addiction: Beta-2 containing nicotinic receptors are involved in nicotine reinforcement but not in withdrawal syndrome. *Psychopharmacology* 2006;187:189–99.

102 Mihalak KB, Carroll FI, Luetje CW. Varenicline is a partial agonist at alpha4beta2 and a full agonist at alpha7 neuronal nicotinic receptors. *Mol Pharmacol* 2006;70:801–5.

3 | The neurobiological mechanisms underlying nicotine dependence

3.1 Introduction

It is now widely accepted that nicotine is the primary addictive component of tobacco smoke. In recent years, however, it has become clear that the psycho-biological mechanisms which mediate the addiction are more complex than they first appeared. This chapter will highlight the ways in which studies with experimental animals have contributed to our understanding of the psychobiological mechanisms which mediate addiction to nicotine and tobacco. The studies provide convincing evidence that nicotine exhibits the principal psychopharmacological properties of a drug of dependence, namely the ability to exhibit reinforcing or rewarding properties and to elicit an abstinence syndrome when the drug is withdrawn precipitously following a period of chronic administration. However, the experimental animal data also indicate that, when compared with many other drugs of dependence, the reinforcing properties of nicotine appear relatively weak. Thus, it may be that nicotine alone does not have the powerful addictive properties necessary to account for the highly addictive nature of tobacco smoking, and that addiction to tobacco reflects complex interactions between nicotine, other stimuli associated with the inhalation of tobacco smoke, and possibly other environmental, social or behavioural stimuli associated with smoking.

3.2 Nicotine self-administration in experimental animals

Many of the studies designed to investigate the psychological and neurobiological factors underlying the rewarding or reinforcing properties of nicotine have employed an intravenous self-administration (IVSA) model in which animals are trained to make a response, commonly to press a lever, to receive a small intravenous dose of nicotine. Thus, in this paradigm, the delivery of nicotine is under the control of the animal and is contingent upon the animal making the appropriate lever-pressing response (Fig 3.1).[1] Although this method of self-administration appears to be very different from the inhalation of nicotine in tobacco smoke, it models the need for the animal to learn and express a specific behaviour in order to receive the drug. The protocol is widely used to explore the reinforcing properties of addictive drugs, and the compounds which serve as reinforcers in this type of protocol (such as cocaine, amphetamine and heroin) are invariably addictive in human beings. Experimental animals do not learn to self-administer saline or non-reinforcing drugs. Furthermore, if saline is substituted for reinforcing drugs, such as nicotine, over a period of three to five sessions the animals stop responding because they learn that the behaviour no longer delivers a rewarding drug. Corrigall and Coen were amongst the first to report that nicotine could serve as a reinforcer in this paradigm in experimental rats, although it has since emerged that De Noble and Mele demonstrated this effect some years previously while working for Philip Morris.[2,3] Subsequent studies have confirmed that the procedure provides a robust means of exploring nicotine-seeking behaviour and the

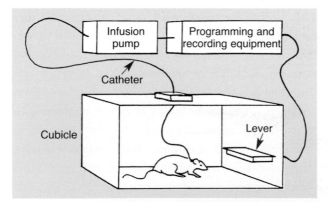

Fig 3.1 Apparatus that can be used to investigate the intravenous self-administration of nicotine in experimental rats. The nicotine is delivered from a syringe pump that is connected via a cannula to a catheter located in the jugular vein. The animals are trained to perform a task (depress a lever in this example) to receive small intravenous injections of nicotine. Reproduced from Stolerman *et al* with permission; copyright Elsevier 1991.[1]

psychological and neurobiological mechanisms which mediate the reinforcing properties of nicotine that are considered pivotal to its ability to cause addiction in humans.[4–6]

3.3 Neurobiology underlying the reinforcing properties of nicotine

Most studies that have sought to establish the neurobiological mechanisms which mediate the reinforcing properties of nicotine have focused on the role of the mesolimbic dopamine (DA) neurons which project to the nucleus accumbens from the ventral tegmental area (VTA) of the midbrain (Fig 3.2). It is widely held that these neurons play a central role in the rewarding or reinforcing properties of drugs of dependence.[7–10] The results reported by Corrigall and colleagues, which showed that selective lesions of the DA-secreting neurons that project to the nucleus accumbens attenuate responding for nicotine by experimental animals, emphasise the importance of these pathways to the reinforcing properties of nicotine.[4] Subsequent studies showed that microinjections of a nicotinic receptor antagonist into the VTA of experimental animals has the same effect. These results

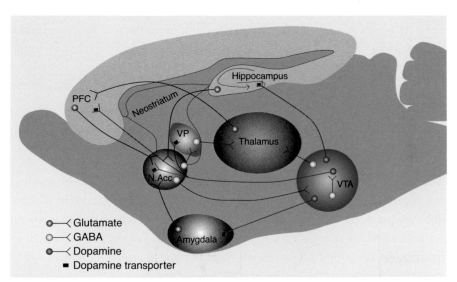

Fig 3.2   Mesolimbic dopamine pathways in the rat brain which project to the forebrain from the ventral tegmental area. The principal pathways project to the nucleus accumbens and prefrontal cortex (PFC). Other neurons project to the amygdala and hippocampus. Other areas of the brain, principally the PFC, amygdala and hippocampus, also send glutamatergic (excitatory) and GABA (inhibitory) projections to the nucleus accumbens. These projections may also exert significant control over dopamine release in the accumbens. Note also the inhibitory GABA feedback loop from the accumbens to the VTA. VTA = ventral tegmental area; N. Acc = nucleus accumbens; PFC = prefrontal cortex.

support the conclusion that the reinforcing properties of nicotine in this paradigm depend upon the stimulation of neuronal nicotine receptors located on or close to the DA cell bodies in the VTA that stimulate the activity (that is, the firing rate or firing pattern) of the DA neurons which innervate the nucleus accumbens.[11]

For the past 20 years or so the effects of treatments which influence the activity of mesolimbic DA neurons have been explored in conscious freely-moving animals using techniques such as *in vivo* microdialysis. Although microdialysis probes are too large to sample DA release directly into the synaptic cleft, they can be used to measure changes in the concentration of DA in the interstitial space between the cells, often referred to as DA overflow. Experiments employing this technique have shown that non-contingent injections of nicotine (injections administered by an experimenter, over which the animal has no control) stimulate mesolimbic DA neurons in a way that results in substantial and sustained increases in the overflow of DA in the interstitial spaces between the cells in the nucleus accumbens.[12,13] More recent imaging studies employing positron emission tomography (PET) have shown that exposure to cigarette smoke also increases DA release in the human nucleus accumbens.[14] This observation supports conclusions drawn from an earlier study with human subjects which suggested that the reinforcing properties of tobacco smoke depend upon the stimulation of DA receptors in the brain.[15] The results imply that studies with experimental animals are able to model changes in DA release in the nucleus accumbens that may be of relevance to our understanding of the neurobiology underlying the dependence upon tobacco in humans.

3.4 The role of the accumbal shell

The nucleus accumbens is composed of two subdivisions: the accumbal core and the accumbal shell, which are anatomically distinct and are thought to have different psychophysiogical roles.[16,17] Anatomically, the accumbal shell seems to form an extension of a clearly limbic structure, the amygdala, and is thought to mediate the rewarding properties of drugs of dependence. This conclusion is strongly supported by the results of a study with another psychostimulant drug of dependence, cocaine, which showed that rats can be trained to self-administer microinjections of the drug directly into the medial shell of the structure whereas they will not learn to self-administer the drug through cannulae targeted at the accumbal core.[18] These results imply that increased DA release in the accumbal shell mediates the reinforcing properties of cocaine.

Sellings and Clarke showed that selective lesions of the DA terminals in the medial shell of the accumbens also attenuate the rewarding properties of amphetamine.[19] These results, when taken together, suggest that the rewarding properties of psychostimulant drugs of dependence, and the acquisition of a response to them, depends specifically upon increased DA overflow in the shell subdivision of the nucleus accumbens. By contrast, selective lesions of the DA projections to accumbal core diminish the increase in locomotion evoked by an injection of amphetamine.[19] This observation supports the anatomical evidence that the accumbal core forms part of the motor systems of the brain and mediates the locomotor stimulant properties of drugs of abuse.

Acute injections of nicotine to nicotine-naive animals increase DA overflow in the medial shell of the accumbens, but have little or no effect in the accumbal core.[20,21] Di Chiara has summarised the evidence that stimulation of DA overflow in the accumbal shell is pivotally important to the development of nicotine-seeking behaviour, and is thus an essential component of the neurobiology underlying the reinforcing properties of the drug.[9,22] In this important sense, it seems to resemble closely the mechanisms which are thought to mediate the reinforcing properties of other psychostimulant drugs of abuse.

Experiments performed some years ago showed that the self-administration of small intravenous injections of cocaine, which were contingent on a response, elicited greater increases in DA overflow in the nucleus accumbens, measured using microdialysis, than that evoked by the non-contingent infusion of matched intravenous doses of cocaine.[23,24] This was a surprising finding because the effect could not be attributed to a simple pharmacological difference in the response between the two groups of rats in the experiment. Recently, Lecca and colleagues have reported that the self-administration of nicotine also results in greater DA overflow in the shell subdivision of the nucleus accumbens than non-contingent administration.[25] Other studies from the same laboratory have shown that the self-administration of cocaine also enhances its effects on DA overflow in the shell subdivision of the nucleus accumbens.[26] These findings highlight an important aspect of the effects of drugs of dependence on this pathway. When the animal has no control over the delivery of an addictive drug (non-contingent drug delivery), the increase in extracellular DA (DA overflow) evoked in the accumbal shell does not exhibit any sensitisation. Indeed, it may exhibit tolerance,[20] resulting in a reduced response to the drug when it is given repetitively as daily injections. However, when an addictive drug is self-administered (that is, when drug delivery is contingent upon the animal making the appropriate response), an enhanced or sensitised DA response to the drug may be observed in the accumbal shell.

Thus, it would appear that the neurobiological mechanisms underlying behavioural conditioning of the response influence the effects that these drugs exert on the DA neurons which innervate the accumbal shell, and that their reinforcing properties are enhanced by the process of self-administration.[25,26] This sensitised response is thought to be important to the development of dependence on drugs such as nicotine, although the neurobiological mechanisms which mediate the effect remain to be established.

3.5 The role of the accumbal core

The acute administration of nicotine to drug-naive rats has little or no effect on DA overflow in the accumbal core. However, if the drug is given repetitively in the form of daily injections, the DA projections become sensitised to the drug and injections of nicotine elicit substantial and sustained increases in DA overflow in this subdivision of the structure.[20,21] Neurons within the accumbal core send major projections to areas of the brain which mediate motor responses and, therefore, it has been proposed that this subdivision of the nucleus accumbens is likely to be involved in the control of locomotion.[16] Because of this, it has been suggested that increased DA overflow in the accumbal core may mediate the locomotor stimulant properties of nicotine,[13,20] which can be measured in animals as an increase in locomotion in a simple activity box. This conclusion is consistent with the evidence that the repetitive administration of nicotine to experimental rats not only results in sensitisation of its effects on DA overflow in the accumbal core, but also in sensitisation of its effects on locomotor activity. This is further supported indirectly by experiments with another psychostimulant, amphetamine, which have shown that selective destruction of DA terminals in the accumbal core of rats attenuate the locomotor stimulant properties of the drug.[19]

Other experiments, however, have shown that expression of the sensitised DA response to nicotine in the core of the accumbens of experimental rats is attenuated or abolished by the prior administration of drugs which block N-methyl-D-aspartic acid (NMDA) receptors for glutamate in the brain.[27,28] Since the microdialysis studies which generated these data could be performed in conscious freely moving animals, it was also possible to test the effects of the antagonists on the sensitised locomotor response to nicotine in the same animals. The experiments showed that the antagonists had no effect on the sensitised locomotor response to nicotine, demonstrating a clear dissociation between the sensitised neural and the behavioural responses to the drug.[27,28] The location of the NMDA receptors remains controversial, although it seems reasonable to suggest that they are located on the DA neurons in the VTA which project to the accumbens (Fig 3.3).[29]

These and other results have led to the suggestion that the sensitised DA responses to nicotine observed in the accumbal core are more likely to be

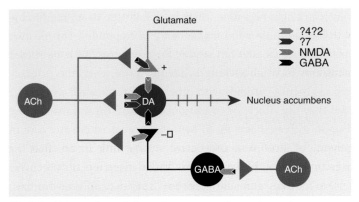

Fig 3.3 Neuronal regulation of mesoaccumbens dopamine neurons within the ventral tegmental area. The neurons shown are thought to synapse with the DA-secreting neurons in the ventral tegmental area (VTA). The ACh (cholinergic neurons) probably project, predominantly, from the pedunculopontine tegmentum (PPTg) in the midbrain. Some of the glutamate-secreting neurons may also project from project from the PPTg, although glutamate neurons in the prefrontal cortex also innervate this structure. The GABA-secreting neurons are located on projections from the nucleus accumbens and on inter-neurons located within the VTA. DA = dopamine; VTA = ventral tegmental area; ACh = acetylcholine.

implicated in the neurobiology underlying nicotine dependence than a simple neurobiological correlate of the locomotor stimulant properties of the drug.[30,31] Interestingly, sensitisation of the DA projections to the accumbal core, evoked by nicotine, appear to be a response which predominates when the drug is given non-contingently by an experimenter rather than when the drug is self-administered. This fact led Lecca and colleagues to argue that the DA projections to the accumbal core do not play a major role in the neurobiology underlying nicotine addiction.[25] This conclusion, however, is controversial. Others have suggested that the sensitised DA responses in the accumbal core, evoked by repetitive administration of the drugs, is fundamentally important to their ability to cause dependence, particularly with regard to the role that conditioned stimuli play in the development of the addiction.[32,33,39] This important issue is considered in the following section.

3.6 The role of conditioned stimuli in nicotine reinforcement

Most of the experiments which have explored nicotine self-administration in experimental animals have incorporated a conditioned stimulus into the experimental design. Indeed, it often proves difficult to demonstrate nicotine self-administration in experimental animals when nicotine is the sole reinforcer.[33,35] In these paradigms, the presentation of nicotine is associated with the presentation of another stimulus (such as a light or a tone which itself has little or no reinforcing properties), and both the acquisition and maintenance of responding for nicotine is facilitated.[35,36] Furthermore, the cues can acquire the properties of a conditioned stimulus which maintains nicotine-seeking behaviour even when the primary reinforcer (nicotine) is absent.[35] Studies with other reinforcers, such as food or cocaine, have shown that the effects of conditioned stimuli on reward-seeking behaviour depend upon increased activity of neurons located within the core of the nucleus accumbens.[37,38] Robinson and Berridge have long argued that the sensitised responses to drugs of abuse mediate the attribution of 'incentive salience' (the capacity to motivate animals to seek the drug) to conditioned stimuli paired with delivery of the drug.[32,33] They argue that these paired stimuli or cues are, therefore, pivotally important to the transition to dependence.

Other animal studies, in which a conditioned stimulus has been paired with intravenous injections of cocaine rather than nicotine, have shown that in the absence of the primary reinforcer (cocaine) the non-contingent presentation of a conditioned stimulus evokes a regionally selective increase in DA overflow in the core of the nucleus accumbens.[39] The presentation of a conditioned stimulus in this way is thought to model the way such stimuli provoke drug-seeking behaviour in abstinent individuals. Thus, the data taken together provide strong support for the conclusion that increased DA overflow in the core subdivision of the nucleus accumbens promotes the compulsive drug-seeking behaviour that characterises the transition to addiction,[34] and that the addicted animal experiences reward from both the drug itself and from the paired stimulus.

3.7 Desensitisation and the neurobiology of addiction

The data summarised in the preceding paragraphs have provided strong support for the conclusion that the reinforcing properties of nicotine, which underpin addiction to the drug, depend upon stimulation of the nicotinic receptors located on the mesolimbic DA projections to the nucleus accumbens. However, if animals are given a second injection of nicotine within 60–90 minutes of the first, the DA response to the drug in the nucleus accumbens is either abolished or attenuated substantially.[40] Also, if nicotine is delivered by continuous infusion using a protocol which maintains the blood nicotine concentration at a level commonly found in the plasma of habitual smokers, the DA response to nicotine injection is blocked.[40,41] Furthermore, the blood nicotine concentration in experimental animals allowed to self-administer nicotine intravenously rises quickly to concentrations that could be expected to desensitise the neuronal nicotinic receptors which mediate its effects on DA overflow in the nucleus accumbens.[34,42] These results, which are supported by other electrophysiological data,[43] suggest that exposure to nicotine not only stimulates the nicotinic receptors in the brain which mediate its effects on DA release in the accumbens, but can also desensitise the receptors if exposure to the drug is sustained even for a relatively short period of time. Many regular smokers smoke in a way that is likely to result in the accumulation of nicotine to a concentration that causes desensitisation of these receptors. Nevertheless, they continue to smoke, and to find the habit reinforcing, when it is unlikely that the nicotine they inhale has any significant effects on DA overflow in the accumbens.[39,44] Thus, hypotheses which seek to explain the addiction to nicotine and tobacco need to explain why self-administration continues at a time when the drug no longer stimulates DA overflow.

3.8 The effects of nicotine withdrawal

There is abundant evidence that many regular smokers experience significant withdrawal effects when they first quit smoking, and that these symptoms can be relieved by nicotine replacement.[45–48] Studies with experimental animals have shown that the abrupt withdrawal of nicotine, following a period of chronic treatment, elicits behavioural changes which are thought to model some of the symptoms of the tobacco abstinence syndrome experienced by abstinent smokers.[49,50] Other behavioural results, using an experimental approach in which rats are trained to stimulate reward pathways within the brain directly through indwelling electrodes, suggest that nicotine withdrawal from experimental animals increases the threshold at which animals find stimuli sufficiently rewarding to elicit a behavioural response.[51,52] This latter response to nicotine withdrawal has been putatively associated with decreased DA overflow in the accumbal shell.[53] The neural responses to nicotine withdrawal are thought to model the neurobiological changes within the brain which mediate the anhedonia (the diminished ability to experience pleasure) experienced by many smokers when they first quit the habit. This symptom may also be experienced following relatively short periods of overnight abstinence. All these behavioural consequences of nicotine withdrawal can be reversed by the administration of nicotine and some non-nicotinic therapies for tobacco dependence such as bupropion.[54,55] Thus, these studies are thought to provide a behavioural model which can be used to explore novel treatments which might be used to alleviate the effects of nicotine withdrawal in humans.

It seems reasonable to suggest that many smokers will seek to maintain their blood nicotine concentrations at levels which prevent the symptoms of withdrawal. This explanation provides one potential reason why smokers may continue to smoke at times when they are unlikely to experience any positive reinforcement derived directly from increased DA overflow in the nucleus accumbens. However, it does not account for the fact that nicotine replacement therapy, although a valuable aid to smoking cessation, is not as efficacious as would be anticipated if tobacco dependence reflected a 'simple' addiction to nicotine.[56] It is important, therefore, to consider more carefully other mechanisms that might be involved, including the putative role of the nicotine delivery vehicle, and of behavioural and other social factors and cues associated with smoking.

When given to animals that have not been preloaded with the drug, a single injection of nicotine results in a sustained increase in DA overflow that persists for 60–120 minutes.[20,21] Mansvelder and colleagues have suggested that the sustained response reflects complex effects on the nicotinic receptors located on the DA neurons themselves and the terminals of GABA and glutamate neurons which innervate the VTA (see Fig 3.3).[57,58] Balfour and colleagues have argued that the sustained DA response to nicotine is related to stimulation of NMDA glutamatergic receptors located on DA-secreting neurons in the VTA, resulting in an increased proportion of the cells which exhibit burst firing.[34,40] Balfour and colleagues also suggest that burst-firing of DA neurons results in DA release from varicosities directly into the interstitial space sampled by the dialysis probes that are commonly used to measure DA overflow in the brain.[40] Common to the hypotheses of both groups is the concept that a single injection of nicotine elicits a sustained increase in extra-cellular DA, which plays a central role in the development of nicotine dependence. Significantly, a study by Lecca and colleagues has shown that the self-administration of nicotine also elicits a sustained increase in DA overflow in the accumbal shell and core which persists for at least 90 minutes into the experimental session.[25]

3.9 The putative role of paracrine dopamine

The presentation of natural rewards, such as food to a hungry rat, also increases DA overflow in the nucleus accumbens. Recent studies, however, suggest that increased DA overflow in the nucleus accumbens is not essential for animals to learn to respond to obtain rewards of this type, although it does facilitate the acquisition of responding and increase its intensity.[50] Thus, increased DA overflow does not seem to mediate the reinforcing properties of natural rewards directly, but to enhance their effects on behaviour. Balfour has argued that the data are consistent with the hypothesis that extracellular DA in the interstitial space between the cells functions as a local hormone which influences the probability that an individual learns the behaviours associated with the delivery of rewards, and also promotes reward-seeking behaviour in response to cues associated with the presentation of the reward (Fig 3.4).[34,40] When compared with natural rewards, the effects of drugs of dependence on the concentration of DA in this extracellular space are large and sustained, and according to this theory it is this property of the drugs which confers on them their addictive potential. The hypothesis, summarised in Fig 3.4, proposes that increased DA overflow in the accumbal shell following nicotine administration confers reinforcing pleasurable properties on behaviours (lever-pressing in rats or smoking in humans), which result in delivery of the drug. It also confers reinforcing properties on the stimuli associated with its delivery, such as the irritation in the mouth, throat and bronchi, the taste of the tobacco smoke, or perhaps even the anticipation of smoking.

It is estimated that a smoker takes eight to 10 puffs on each cigarette which may be inhaled during periods when extracellular DA in the accumbal shell has been elevated by the nicotine boli in the first two or three puffs. Thus, the repetitive nature of cigarette smoking lends itself particularly well to rapid

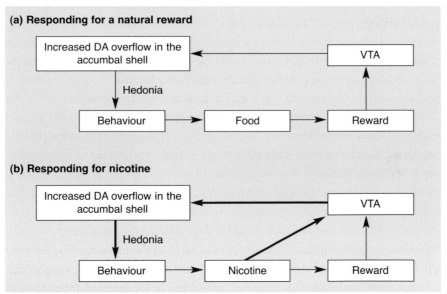

Fig 3.4 The role of increased dopamine overflow in the accumbal medial shell on responding for a reward. **(a)** The proposed circuitry by which a natural reward, such as food, increases DA overflow in the accumbal medial shell. It is hypothesised that this increased DA overflow confers rewarding or hedonic characteristics on a behaviour, such as a lever-pressing response, that results in presentation of the reward.
(b) Nicotine may greatly enhance the hedonia associated with the behaviour by directly increasing extracellular DA in the accumbal shell, bypassing the need for the drug itself to exhibit rewarding properties. DA = dopamine. Reproduced from Balfour *et al* with permission; copyright Elsevier 2000.[40]

acquisition of the addiction of tobacco. The hypothesis summarised in Fig 3.4 relates specifically to conditioned stimuli closely associated with the inhalation of tobacco smoke. However, it is also possible that increased DA overflow in the accumbal shell mediates the 'pleasure' smokers derive from environmental stimuli, such as specific locations or activities like taking a break from looking after the children, having a drink in a pub or relaxing after a meal, which they associate with smoking.

The hypothesis also proposes that extracellular DA in the accumbal core acts as a local hormone. It is proposed that its role in this subdivision of the nucleus accumbens is to increase the probability that the stimuli associated with smoking promote the desire or craving to smoke. Thus, when nicotine itself increases DA overflow in this part of the accumbens, it may serve to increase the craving to continue smoking. If a smoker experiences this effect by smoking following a period of abstinence, the hypothesis predicts that it will provide one of the powerful drivers which promote relapse. Furthermore, it seems reasonable to suggest that the regionally selective increases in DA overflow in the accumbal core, evoked by non-contingent exposure to conditioned stimuli or cues associated with delivery of an addictive drug, also contribute significantly to the cravings which commonly result in relapse.[34,39,60]

As indicated earlier in this review, nicotine alone is a relatively weak reinforcer in a self-administration paradigm, and robust self-administration of the drug is best achieved using protocols which incorporate a conditioned reinforcer.[35,36] The initial studies which demonstrated the facilitatory role of conditioned stimuli employed small intravenous injections of nicotine that were contingent upon the animal making a response (pressing a lever) in order to receive the injections of nicotine. It was assumed that for nicotine to confer reinforcing properties on a conditioned stimulus, such as a simple light, it was necessary for the nicotine injections paired with the stimulus to be contingent upon a response; that is, self-administered in a manner which could be controlled by the experimental animal. More recent studies suggest that this may not be true. Donny and colleagues have shown that non-contingent injections of nicotine, which are not under the control of the animals, also have the ability to confer significant reinforcing properties on a light stimulus which are equal in magnitude to those seen in animals which self-administer the drug.[61]

This observation may be profoundly significant because it suggests that, in the self-administration paradigms employing animal models, responding is reinforced to a significant extent by the conditioned stimulus rather than the drug directly.[60,61] In this theory, an important role of nicotine in developing dependence is to greatly enhance the reinforcing or rewarding properties of stimuli paired with delivery of the drug. If this observation translates to tobacco smoke, it implies that people may smoke, predominantly, for conditioned stimuli associated with inhaling tobacco smoke, rather than a simple dependence on the boli of nicotine present in each puff. This conclusion gains much support from the studies of Rose and colleagues who have shown that sensory stimuli, such as the taste of the tobacco smoke or the irritation of the mouth and bronchi experienced by a smoker when they inhale tobacco smoke, play a pivotal role in regulating smoking behaviour and the craving to smoke.[62] Furthermore, these stimuli seem to be particularly important for more highly addicted smokers.[63] Balfour has argued that these sensory stimuli serve as powerful conditioned reinforcers for smokers.[34,60] The evidence that nicotine also plays a pivotal role in the addiction to tobacco is, nevertheless, compelling. Thus, if the hypothesis that sensory stimuli present in tobacco smoke are central to development of addiction to tobacco smoke is correct, it remains necessary to explain the role of nicotine. Balfour has sought to explain this paradox by suggesting that nicotine evokes changes in the brain which confer or magnify the reinforcing properties of sensory stimuli such as a light in animal studies or the sensory stimuli in tobacco smoke which would otherwise lack significant efficacy as a reinforcer.[34,60]

3.10 A unifying hypothesis

There is growing consensus that increases in extracellular DA in the accumbal shell and core, evoked by nicotine, play complementary roles in the development of addiction to tobacco. This hypothesis proposes that the primary role of extracellular DA in the medial shell of the nucleus accumbens is to influence the probability that an individual exhibits a particular behaviour (Fig 3.5).[40] Increased extracellular DA in the accumbal shell is posited to enhance the 'rewarding' or hedonic properties of a behaviour, thereby increasing the probability that it is repeated. The result of this effect is to facilitate the acquisition and maintenance of behaviours that generate rewards so that they are learned efficiently. Responding for conditioned stimuli, associated with the presentation of either natural or drug rewards, seems to depend upon increased neural activity within the accumbal core.[38,64] Non-contingent presentation of a conditioned stimulus, previously paired with an addictive drug, causes a regionally selective increase in DA overflow in the accumbal core.[39] The hypothesis predicts that increased DA overflow in this subdivision of the accumbens serves to increase the probability that exposure to a conditioned stimulus, associated with delivery of a reward, elicits Pavlovian reward-seeking behaviour. When this phylogenetically old, and relatively primitive, mechanism is stimulated

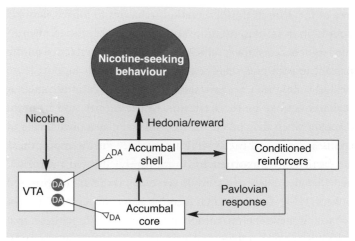

Fig 3.5 The putative roles of hedonia and conditioned reinforcers in nicotine-seeking behaviour. The mechanisms proposed in this review suggest that increased DA overflow in the medial shell and core of the nucleus accumbens, evoked by an injection of nicotine, play complementary roles in the expression of nicotine-seeking behaviour. The hypothesis posits that, in both subdivisions of the accumbens, extracellular DA serves to promote or amplify the signals that project from or through the structure. Stimulation of the projections to the medial shell of the accumbens enhances the hedonic value of the behaviour itself and of sensory and environmental stimuli associated with the delivery of nicotine. Stimulation of the projections to accumbal core promotes the effects of conditioned reinforcers or stimuli on nicotine-seeking behaviour. These conditioned responses can be amplified further by stimulation of the DA projections to the medial shell, through which neurones from the core project. DA = dopamine. Reproduced from Balfour *et al* with permission; copyright Elsevier 2000.[40]

by drugs that elicit the substantial and sustained increases in DA overflow in the accumbal core and shell, they stimulate a powerful craving for the drug and confer significant reinforcing properties on the conditioned stimuli. As a result, even in the absence of the primary drug reinforcer, addicted individuals will continue to respond for the conditioned stimulus alone. Furthermore, the non-contingent presentation of the stimulus evokes drug-seeking behaviour.

The merit of the hypothesis lies in its ability to explain why a relatively weakly reinforcing drug, nicotine, can result in a powerful addiction to tobacco. Central to the hypothesis is the concept that neuroadaptive changes in the regulation of mesolimbic DA neurons result in the compulsive drug-seeking behaviour that characterises the addiction to tobacco. The hypothesis predicts that the substantial and sustained increases in DA overflow evoked by nicotine in the accumbal shell confer powerful reinforcing properties on behaviours that deliver the drug, such as lever-pressing in experimental rats or smoking in humans, and also on stimuli associated with its delivery.[34,60] Its effects on drug-seeking behaviour lie in its ability to act directly upon pathways within the brain which mediate behaviour and to evoke in these pathways unphysiologically large and sustained increases in extracellular DA. As a result, this learned or conditioned behaviour comes to dominate the behavioural repertoire of the individual.

For much of the day, the nicotine in tobacco smoke may not elicit the increases in DA overflow which the studies cited above imply mediate the dependence upon tobacco because receptors which mediate the responses are desensitised.[40] During these periods, it seems likely that smoking is reinforced solely by the conditioned stimuli present in tobacco smoke experienced by the smoker.[34,40] Most smokers, however, experience times of abstinence each day, when they are asleep, for example, during which the receptors are resensitised. Thus, each day, there will also be occasions when the nicotine present in tobacco smoke again elicits the increases in DA overflow which are thought to be fundamentally important to its ability to confer potent reinforcing properties on stimuli present in the smoke, thereby maintaining the powerful addictive nature of the habit. In essence, therefore, tobacco smoking could be seen as a form of second order schedule of reinforcement in which responding is only occasionally reinforced by the direct effects of nicotine in the brain. Importantly, and perhaps surprisingly, the hypothesis does not require that the drug itself has euphoriant or rewarding properties in its own right. Indeed, there is some evidence that nicotine can have aversive properties which also depend upon its ability to stimulate DA neurons in the ventral tegmental area.[65,66] These authors suggest that the data currently available support the conclusion that increased DA release in the nucleus accumbens may not mediate directly a rewarding property of nicotine itself, but has an important role in facilitating the acquisition and maintenance of specific goal-directed behaviours (lever-pressing in rats, smoking in humans) that result in presentation of the drug or stimuli.[65,66] This conclusion is entirely consistent with the working hypothesis presented here, in which increased DA overflow in the accumbal shell and core play complementary roles in the promotion of behaviours that deliver nicotine and, especially, the conditioned stimuli associated with delivery of nicotine.

3.11 Inhibition of monoamine oxidase

It is also important to consider the possibility that tobacco smoke may contain other compounds that are reinforcing in their own right or which could potentiate the addictive properties of nicotine through a pharmacological interaction, rather than a psychological or behavioural mechanism. Fowler and colleagues have reported that tobacco smoke contains a compound, not nicotine itself, which inhibits the enzyme, monoamine oxidase (MAO).[67,68] This is the main enzyme responsible for the metabolism of DA in the human brain. They have argued that one of the primary consequences of this enzyme inhibition will be the potentiation of the effects of nicotine on monoamine, particularly DA, release in the brain, thereby enhancing the addictive potential of the nicotine inhaled in tobacco smoke. This conclusion is supported by the results of studies with experimental rats which have shown that pharmacological inhibition of MAO in the brain does enhance the reinforcing properties of nicotine when they are assessed using a self-administration paradigm.[69]

3.12 Conclusions

▸ Drug dependence has commonly been perceived primarily as a 'drug-seeking' behaviour in which an addicted individual exhibits a powerful desire or craving for the drug, and that on each occasion the addicted individual takes the drug, he or she experiences a powerful rewarding effect which is mediated directly by the drug itself.

▸ However, evidence from both animal and human studies indicates that addiction to tobacco smoking is more complex. It also seems to depend critically upon the ability of nicotine to confer powerful rewarding properties on other sensory cues arising from the process of smoking, or possibly also the circumstances in which smoking occurs.

▸ The direct reinforcing properties of nicotine are experienced predominantly only after periods of temporary abstinence – after sleep, for example. For much of the remainder of the smoking day, nicotinic receptors are desensitised and the nicotine inhaled by the smoker probably does not cause stimulation of the pathways, particularly the mesolimbic DA system, implicated in the development of dependence. During these periods, smoking is probably reinforced primarily by conditioned stimuli (sensory cues) present in the smoke.

▸ By smoking in this way, smokers can maintain their blood nicotine level at a concentration sufficient to prevent the aversive consequences of withdrawal, while continuing to derive some positive reinforcement from the conditioned stimuli present in the smoke, and possibly other behavioural cues associated with smoking.

▸ Hypotheses concerning the neurobiological mechanisms have been derived predominantly from studies with experimental animals. As far as it has been possible to test them, the pharmacological responses to nicotine inhaled by humans in tobacco smoke elicit similar effects in the human brain to those observed in experimental animals. It is important to acknowledge, however, that the rats and mice commonly used for experimental studies lack the complex cognitive skills and social behaviour of humans. It is also important to acknowledge that the interplay between the pharmacological properties of nicotine inhaled in tobacco smoke and the cues and conditioned stimuli associated with human tobacco dependence are likely to be even more complex than those revealed by animal studies. Future studies which seek to establish better treatments for tobacco addiction might usefully focus on this aspect of the problem.

References

1   Stolerman IP, Shoaib M. The neurobiology of tobacco addiction. *Trends Pharmacol Sci* 1991;12:467–73.

2   Corrigall WA, Coen KM. Nicotine maintains robust self-administration in rats on a limited access schedule. *Psychopharmacology* 1989;99:473–8.

3   De Noble VJ, Mele PC. Intravenous nicotine self-administration in rats: effects of mecamylamine, hexamethonium and naloxone. *Psychopharmacology* 2006;184:266–72.

4   Corrigall WA, Franklin KJB, Coen KM, Clarke PBS. The mesolimbic dopaminergic system is implicated in the reinforcing effects of nicotine *Psychopharmacology* 1992;107:285–9.

5   Donny EC, Caggiula AR, Knoof S, Brown C. Nicotine self-administration in rats. *Psychopharmacology* 1995;122:390–4.

6   Shoaib M, Schindler CW, Goldberg SR. Nicotine self-administration in rats: strain and nicotine pre-exposure effects on acquisition. *Psychopharmacology* 1997;129:35–43.

7   Di Chiara G, Imperato A. Drugs abused by humans preferentially increase synaptic dopamine concentrations in the mesolimbic system of freely moving rats. *Proc Nat Acad Sci* 1988;85:5274–8.

8   Di Chiara G. Drug addiction as a dopamine-dependent associative learning disorder. *Eur J Pharmacol* 1999;375:13–30.

9   Di Chiara G. Behavioural pharmacology and neurobiology of nicotine reward and dependence In: Clementi C, Fornasari D, Gotti C (eds), *Handbook of experimental pharmacology vol 14*: 603–750. Berlin: Springer, 2000.

10   Wise RA. Dopamine, learning and motivation. *Nat Rev Neurosci* 2004;5:483–94.

11   Corrigall WA, Coen KM, Adamson KL. Self-administered nicotine activates the mesolimbic dopamine system through the ventral tegmental area. *Brain Res* 1994;653:278–84.

12   Imperato A, Mulas A, Di Chiara G. Nicotine preferentially stimulates dopamine release in the limbic system of freely moving rats. *Eur J Pharmacol* 1986;132:337–8.

13   Benwell MEM, Balfour DJK. The effects of acute and repeated nicotine treatment on nucleus accumbens dopamine and locomotor activity. *Br J Pharmacol* 1992;105:849–856.

14   Dagher A, Bleicher C, Astin JA *et al.* Reduced dopamine D1 receptor binding in the ventral striatum of cigarette smokers. *Synapse* 2001;42:48–53.

15   Dawe S, Gerada C, Russell MAH, Gray JA. Nicotine intake in smokers increase following a single dose of haloperidol. *Psychopharmacology* 1995;117:110–5.

16   Heimer L, Zahm DS, Churchill L, Kalivas PW, Wohltman C. Specificity in the projection patterns of accumbal core and medial shell in the rat. *Neuroscience* 1991;41:89–125.

17   Zahm DS, Brog JS. On the significance of subterritories in the 'accumbens' part of the rat ventral striatum. *Neuroscience* 1992;50:751–67.

18   Rodd-Henricks ZA, McKenzie DL, Ting-Kai L, Murphy J M, McBride WJ. Cocaine is self-administered into the medial shell but not the core of the nucleus accumbens of Wistar rats. *J Pharmacol Exp Ther* 2002;303:1216–26.

19   Sellings LHL, Clarke PBS. Segregation of amphetamine reward and locomotor stimulation between nucleus accumbens medial shell and core. *J Neurosci* 2003;23:6295–303.

20   Cadoni C, Di Chiara G. Differential changes in the accumbens medial shell and core dopamine in behavioural sensitization to nicotine. *Eur J Pharmacol* 2000;387:R23–R25.

21   Iyaniwura TT, Wright AE, Balfour DJK. Evidence that mesoaccumbens dopamine and locomotor responses to nicotine in the rat are influenced by pre-treatment dose and strain. *Psychopharmacology* 2001;158:73–9.

22   Di Chiara G. Role of dopamine in the behavioural actions of nicotine related to addiction. *Eur J Pharmacol* 2000;393:295–314.

23   Di Ciano P, Blaha CD, Phillips AG. Changes in dopamine oxidation currents in the nucleus accumbens during unlimited-access self-administration of d-amphetamine by rats. *Behav Pharmacol* 1996;7:714–29.

24   Hemby SE, No C, Koves TR, Smith JE, Dworkin SI. Differences in extracellular dopamine concentration in the nucleus accumbens during response-dependent and response-independent cocaine administration in the rat. *Psychopharmacology* 1997;133:7–16.

25   Lecca D, Cacciapaglia F, Valentini V, Gronli J, Spiga S, Di Chiara G. Preferential increase of extracellular dopamine in the rat nucleus accumbens shell as compared to that in the core during acquisition and maintenance of intravenous nicotine self-administration. *Psychopharmacology* 2006;184:435–46.

26   Lecca D, Cacciapaglia F, Valentini V, Acquas E, Di Chiara G. Differential neurochemical and behavioral adaptation to cocaine after response contingent and noncontingent exposure in the rat. *Psychopharmacology* 2007;191:653–67.

27   Shoaib M, Benwell MEM, Akbar MT, Stolerman IP, Balfour DJK. Behavioural and neurochemical adaptations to nicotine in rats: influence of NMDA antagonists. *Br J Pharmacol* 1994;111:1073–80.

28   Balfour DJK, Birrell CE, Moran RJ, Benwell MEM. Effects of acute D-CPPene on meso-accumbens dopamine responses to nicotine in the rat. *Eur J Pharmacol* 1996;316:153–6.

29   Schilström B, Nomikos GG, Nisell M, Hertel P, Svensson TH. N-methyl-D-aspartate receptor antagonisms in the ventral tegmental area diminishes the systemic nicotine-induced dopamine release in the nucleus accumbens. *Neuroscience* 1998;82:781–9.

30   Benwell MEM, Holtom PE, Moran RJ, Balfour DJK. Neurochemical and behavioural interactions between ibogaine and nicotine in the rat. *Br J Pharmacol* 1996;117:743–9.

31   Balfour DJK, Benwell MEM, Birrell CE, Kelly RJ, Al-Aloul M. Sensitization of the meso-accumbens dopamine response to nicotine. *Pharmacol Biochem Behav* 1998;59:1021–30.

32   Robinson TE, Berridge KC. The neural basis of drug craving: an incentive-sensitization theory of addiction. *Brain Res Rev* 1993;18:247–91.

33   Robinson TE, Berridge KC. Addiction. *Ann Rev Psychol* 2003;54:25–53.

34   Balfour DJK. The neurobiology of tobacco dependence: a preclinical perspective on the role of the nucleus accumbens. *Nic Tob Res* 2004;6:899–912.

35   Caggiula AR, Donny EC, White AR, Chaudhri N, Booth S, Gharaib MA, Hoffman A, Perkins KA, Sved AF. Cue dependency of nicotine self-administration and smoking. *Pharmacol Biochem Behav* 2001;70:515–30.

36   Caggiula AR, Donny EC, Chaudhri N. Importance of nonpharmacological factors in nicotine self-administration. *Physiol Behav* 2002;*77*:683–7.

37   Parkinson JA, Olmstead MC, Burns LH, Robbins TW, Everitt BJ. Dissociation in effects of lesions of the nucleus accumbens core and shell on appetitive pavlovian approach behaviour and the potentiation of conditioned reinforcement and locomotor activity by D-amphetamine. *J Neurosci* 1999;19:2401–11.

38   Ito R, Robbins TW, Everitt BJ. Differential control over cocaine-seeking behaviour by nucleus accumbens core and shell. *Nature Neurosci* 2004;*7*:389–97.

39   Ito R, Dalley JW, Howes SR, Robbins TW, Everitt BJ. Dissociation in conditioned dopamine release in the nucleus accumbens core and medial shell in response to cocaine cues and during cocaine-seeking behaviour in rats. *J Neurosci* 2000;*20*:7489–95.

40   Balfour DJ, Wright AE, Benwell MEM, Birrell CE. The putative role of extra-synaptic mesolimbic dopamine in the neurobiology of nicotine dependence. *Behav Brain Res* 2000;113:73–83.

41   Benwell MEM, Balfour DJK Birrell CE. Desensitisation of nicotine-induced dopamine responses during constant infusion with nicotine. *Br J Pharmacol* 1995;114:211–17.

43   Pidoplichko V, De Biasi M, Williams JT, Dani J. Nicotine activates and desensitizes midbrain dopamine neurons. *Nature* 1997;390:401–4.

42   Shoaib M, Stolerman IP. Plasma nicotine and cotinine levels following intravenous nicotine self-administration in rats. *Psychopharmacology* 1999;143:318–21.

44   Cadoni C, Solinas M, Di Chiara G. Psychostimulant sensitization: differential changes in accumbal medial shell and core dopamine. *Eur J Pharmacol* 2000;388:69–76.

45   West RJ. Nicotine: a dependence-producing substance. *Prog Clin Biol Res* 1988;261:237–59.

46   Glassman AH, Helzer JE, Covey LS *et al*. Smoking, smoking cessation and major depression. *J Am Med Assoc* 1990;264:1546–9.

47   Hughes JR, Gust SW, Skoog K, Keenan RM, Fenwick JW. Symptoms of tobacco withdrawal A replication and extension. *Arch Gen Psychiat* 1991;48:52–9.

48   West R, Shiffman S. Effect of oral nicotine dosing forms on cigarette withdrawal symptoms and craving: a systematic review. *Psychopharmacology* 2001;155:115–122.

49   Malin DH, Lake JR, Newlin-Maultsby P *et al*. Rodent model of nicotine abstinence syndrome. *Pharmacol Biochem Behav* 1992;43:779–84.

50   Malin DH. Nicotine dependence studies with a laboratory model. *Pharmacol Biochem Behav* 2001;70:551–9.

51   Epping-Jordan MP, Watkins SS, Koob GF, Markou A. Dramatic decreases in brain reward function during nicotine withdrawal. *Nature* 1998;*393*:76–9.

52   Kenny PJ, Markou A. Neurobiology of the nicotine withdrawal syndrome. *Pharmacol Biochem Behav* 2001;70:531–49.

53   Hildebrand BE, Nomikos GG, Hertel P, Schilström B, Svensson TH. Reduced dopamine output in the nucleus accumbens but not the prefrontal cortex in rats displaying mecamylamine-precipitated nicotine withdrawal syndrome. *Brain Res* 1998;779:214–225.

54   Cryan JF, Bruijnzeel AW, Skjei KL, Markou A. Bupropion enhances brain reward function and reverses the affective and somatic aspects of nicotine withdrawal in the rat. *Psychopharmacology* 2003;168:347–58.

55   Malin DH , Lake JR, Smith TD *et al*. Bupropion attenuates nicotine abstinence syndrome in the rat. *Psychopharmacology* 2006;184:494–503.

56   Balfour DJK, Fagerström KO. Pharmacology of nicotine and its therapeutic use in smoking cessation and neurodegenerative disorders. *Pharmacol Ther* 1996;72:51–81.

57   Mansvelder HD, McGehee DS. Long-term potentiation of excitatory inputs to brain
      reward areas by nicotine. *Neuron* 2000;27:349–57.

58   Mansvelder HD, Keath, JR, McGehee DS. Synaptic mechanisms underlie nicotine-induced
      excitability of brain reward areas. *Neuron* 2002; 33:905–19.

59   Cannon CM, Palmiter RD. Reward without dopamine. *J Neurosci* 2003;23:10827–31.

60   Balfour DJK. Complementary roles for the accumbal shell and core in nicotine
      dependence. In: Corrigall WA (ed), *Understanding nicotine and tobacco addiction.* Novartis
      Symposium, in press.

61   Donny EC, Chaudhri N, Caggiula AR *et al.* Operant responding for a visual reinforcer in
      rats is enhanced by noncontingent nicotine: implications for nicotine self-administration
      and reinforcement. *Psychopharmacology* 2003;169:68–76.

62   Rose JE, Behm FM, Levin ED. Role of nicotine dose and sensory cues in the regulation of
      smoke intake. *Pharmacol Biochem Behav* 1993;44:891–900.

63   Rose JE, Behm FM, Westman EC, Johnson M. Dissociating nicotine and nonnicotine
      components of cigarette smoking. *Pharmacol Biochem Behav* 2000;67:71–81.

64   Hall J, Parkinson JA, Connor TMF, Dickinson A, Everitt BJ. Involvement of the central
      nucleus of the amygdala and nucleus accumbens core in mediating Pavlovian influences
      on instrumental behaviour. *Eur J Neurosci* 2001;13:1984–92.

65   Laviolette SR, Alexson TO, van der Krooy D. Lesions of the tegmental pedunculopontine
      nucleus block the rewarding effects and reveal the aversive effects of nicotine in the ventral
      tegmental area. *J Neurosci* 2002;22:8653–60.

66   Laviolette SR, van der Krooy D. The neurobiology of nicotine addiction: bridging the gap
      from molecules to behaviour. *Nat Rev Neurosci* 2004;5:55–65.

67   Fowler JS, Logan J, Wang GJ, Volkow ND. Monoamine oxidase and cigarette smoking.
      Neurotoxicology 2003;24:75–82.

68   Fowler JS, Volkow ND, Wang GJ *et al.* Brain monoamine oxidase A inhibition in cigarette
      smokers. *Proc Natl Acad Sci USA* 1996;93:14065–9.

69   Guillem K, Vouillac C, Azar MR *et al.* Monoamine oxidase inhibition dramatically
      increases the motivation to self-administer nicotine in rats. *J Neurosci;* 25:8593–600.

4 | Mechanisms of tobacco addiction in humans

4.1 Background and perspective

As outlined in the previous chapter, research into the mechanisms of addiction in animal models in recent years has advanced understanding of the likely process and course of tobacco addiction in humans, and has implications for tobacco disease control and policy. It is now well established that nicotine is the drug in tobacco that is predominantly, but not entirely, responsible for the strength and course of addiction in tobacco smokers and the maintenance of tobacco smoking in the population. A report published by the Royal College of Physicians in 2000 described the reasons why cigarettes are an especially addictive form of nicotine delivery, and how nicotine addiction from tobacco smoking compared with other forms of drug addiction.[1] The report also described the biological mechanisms of addiction in humans, and how tobacco products are designed, manufactured and marketed to increase addiction risk.[1] Since then, the evidence base supporting understanding of nicotine as an addictive drug has continued to expand. Additionally, the importance of the cigarette as a vehicle to enhance the addictive effects of nicotine, and of neural mechanisms and genetic modulators of nicotine addiction, is becoming increasingly well documented.

4.2 The addictiveness of tobacco products

All leading medical authorities that have reviewed the evidence, including the Royal College of Physicians, the US Surgeon General, the US Food and Drug Administration (FDA), and the World Health Organization,[1-4] have concluded that tobacco products are highly addictive. In comparison with addiction to opioids, stimulants, sedatives, alcohol or other so-called 'classic' addicting substances, the risk of the development of addiction following initial use, the severity of the addiction, the risk of adverse health consequences and persistence of use even in the face of harm are all especially high with tobacco.[1–3] Tobacco products optimise the delivery to the user of addictive doses of nicotine. The speed of absorption and addictive impact of nicotine is manipulated and optimised by the designs and ingredients used to make tobacco products.[4,5] Compared with other addicting substances (including cocaine and heroin), initial use of nicotine is more likely to lead to addictive use, and the prevalence of addiction among all users is higher than that observed for other addictive substances. However, physiological dependence, intoxication and withdrawal are less pronounced than is the case with opioids and sedatives.[1,2,6] In contrast with sedatives such as alcohol, nicotine intoxication occurs rarely in regular tobacco users.[1,2]

4.3 The role of nicotine as the addictive drug in tobacco

Tobacco contains hundreds of substances; tobacco smoke contains more than 4,000, including several with potential behavioural effects. Nicotine is common to all tobacco products – smoked and smokeless.[2,7,8] The addiction potential of nicotine is enhanced by the deliberate engineering of tobacco products as vehicles for nicotine delivery, among which cigarettes are recognised as the most addictive of all nicotine delivery systems.[1] These facts have long been recognised by the tobacco industry,[2–4,9–11] as illustrated in a summary conclusion by a lead Philip Morris nicotine researcher, William L Dunn:

> *The cigarette should be conceived not as a product but as a package. The product is nicotine .... Think of the cigarette pack as a storage container for a day's supply of nicotine .... Think of the cigarette as a dispenser of a dose unit of nicotine .... Think of a puff of smoke as the vehicle of nicotine .... Smoke is beyond question the most optimized vehicle of nicotine and the cigarette the most optimized dispenser of smoke.*[10]

The chemistry and pharmacology of nicotine alone qualify it as a potent and powerfully addicting drug.[1,2] Specifically, nicotine meets all established criteria for a drug that produces addiction or, more technically, dependence and withdrawal (upon its abrupt abstinence) in the animal models used for making such determinations and investigating the mechanisms of addicting effects (see Chapter 3). Tobacco products are designed to enable nicotine to be readily and rapidly extracted, absorbed, and distributed to the central nervous system. The drug is five to ten times more potent than cocaine or morphine in producing behavioural and psychic effects associated with addiction potential in humans, including measures of pleasure and liking.[2] Nicotine administration modulates levels of dopamine and other neurotransmitters that mediate addictive drug effects and primary biological drives such as reinforcement by food and other factors. Animal models demonstrate that nicotine produces reinforcing effects, tolerance, physical dependence and withdrawal, and suggest that some effects can be long lasting, including receptor up-regulation and a propensity to relapse to use following abstinence.[1, 2]

Some effects of nicotine, such as dose-related activation and inactivation of receptors and reversible tolerance were studied more than a century ago by pioneering researchers such as Langley and Dixon, while effects on specific nicotinic receptor subpopulations such as the α7 that appear to mediate reinforcement were not conclusively demonstrated until the 1990s.[12–14] It has been recognised for centuries, and highlighted in the 1988 US Surgeon General's report on nicotine addiction, that nicotine can also contribute to the regulation of mood, cognition, attention, body weight and other functions that undoubtedly contribute to the power of the addiction and difficulty in giving up all forms of nicotine. An apparent genetic vulnerability to tobacco addiction and genetically conferred differences in nicotine addiction liability have been suspected, but not well understood, since at least the late 20th century.[2,15,16]

4.4 The cycle of human tobacco use and addiction

4.4.1 Dependence and withdrawal

Dependence and withdrawal are separate disorders related to tobacco use. Each can be diagnosed by criteria issued by the World Health Organization in its International Classification of Diseases (ICD 10) and according to generally similar criteria by the American Psychiatric Association in its Diagnostic and Statistical Manual (DSM-IV).[17,18] Essentially, dependence is the disorder of repetitive and compulsive use of a drug, whereas withdrawal is a syndrome of signs and symptoms that is precipitated when a regular user of a drug abruptly abstains. The more general term 'addiction' is often used synonymously with dependence by these same organisations, but is more variably defined. In this report, the term addiction is used interchangeably with dependence.

Dependence

The specific criteria for drug dependence used by both ICD 10 and DSM-IV are broadly similar, including phenomena such as the development of tolerance to the effect of the drug, the occurrence of withdrawal symptoms on discontinuing use, increased drug use over time, compulsive use despite efforts to reduce or control ('out-of-control') drug use, and use in the face of adverse consequences. Whether the drug in question is cocaine, heroin, alcohol or nicotine, the criteria are the same, although drugs differ somewhat in their profiles of these phenomena. Intoxication is not a criterion of dependence, but is a frequent consequence of the use of many addictive drugs, and this too varies widely between drugs, occurring most prominently with alcohol and sedatives, and least commonly with stimulants and nicotine.[1]

It is possible to use tobacco products occasionally without developing addiction, though even very low levels of tobacco exposure can be harmful to health. For example, second-hand smoke significantly increases the risk of lung cancer and many other diseases.[1,19,20] With respect to most addictive drugs, most people who try the drug do not become addicted. For example, although cocaine is widely accepted as a highly addictive drug, several studies have shown that only about 5–6% of people who try cocaine develop dependence within two years of first use.[21] The cigarette form of tobacco appears to be the most likely to lead to regular use and addiction in Britain and many other countries.[1,20] Among young people who try smoking cigarettes, between one third and one half are currently likely to become regular smokers within two to three years. Many of these will try and fail to quit during their youth, but their addiction will drive them to persist in their use for many years, with drastic consequences to their health and life expectancy, and costs to healthcare, families and employers.[1,22] The escalation from initial use to dependence is variable between individuals and is modulated by a variety of factors including opportunities to use tobacco, cost of the product, perception of harmfulness, social attitudes, and smoking behaviour of peers and family members. Several studies have found that the transition from initial use to regular use (variably defined as 'daily', 'five times per week', or '100 cigarettes') takes two to three years among adolescents.[23] Fewer studies are available in which measures of dependence have been collected from youth to determine the trajectory of development of dependence. However, the evidence suggests that at least some measures (trying and failing to quit, having strong cravings and having difficulty concentrating without cigarettes, for example) can emerge within weeks of initial use.[24]

The most important determinant of the initial brand of cigarettes used appears to be the marketing and imagery associated with it, and many people remain loyal to that same brand for many years. While it appears that nicotine dosing characteristics can be manipulated by altering the ease of extraction of nicotine and the fraction of nicotine that is in the more bioavailable free-base (unprotonated) form, virtually any cigarette can be smoked to deliver a range of doses of nicotine.

Manipulation of the free-base fraction of nicotine with ammonia compounds and other ingredients may also alter the addictiveness of the products, but because most inhaled nicotine is absorbed, the extent to which such manipulation affects the graduation from initial use to addiction is unclear.[1,3,25]

However, nicotine dosing characteristics are not the only factors in attracting new smokers and facilitating the graduation to regular use and addiction. Ingredients and design features that make cigarettes more palatable and attractive to target new smokers include a wide range of flavourings, smoke modifiers, burn accelerants and filter ventilation techniques.[4,7,11,26–29]

Control of the fraction of free-base nicotine seems to be more important for oral smokeless tobacco products (chewing tobacco, moist snuff and snus), in making the product attractive to particular populations, than for cigarettes, because the speed and dose of nicotine delivered by these products depends more strongly on the free-base fraction of nicotine.[3,30–32] This is because absorption of nicotine from oral smokeless tobacco products occurs predominantly in the mouth. Free-base nicotine is rapidly absorbed orally into systemic circulation by transfer across the buccal mucosa into the oral capillary bed, whereas protonated nicotine is much more slowly absorbed.[3,30–32] Thus, to facilitate the development of addiction and minimise effects such as nausea that might discourage repeated use, smokeless tobacco companies have developed and marketed products that vary widely in their nicotine delivery characteristics. Products aimed at new tobacco users (termed 'starter' products within the industry) are designed to deliver their nicotine more slowly and in smaller amounts than products marketed to experienced, and presumed addicted and tolerant, tobacco users.[3,23,30,32,33]

In addition to manipulating nicotine dosing characteristics to promote and maintain addiction, factors such as palatability and attractiveness also increase the likelihood that target populations will try, and continue to use, smokeless products.[3] For example, in recent years, there has been a proliferation of smokeless tobacco products designed to be more convenient to use and to minimise the necessity of spitting, with unit doses of tobacco prepared in small sachets and lozenge forms, and in more user-friendly packaging such as pop-open containers (see Chapter 5).[34]

Comparing the risk of escalation from initial use to addiction across different tobacco product types is difficult. Although there is an extensive base of epidemiological and laboratory evidence for cigarette smoking, the epidemiology and natural history of patterns of use of pipes, cigars, and smokeless tobacco is much less well documented. Moreover, daily intake for these products is less well defined than for cigarettes. Nonetheless, it appears likely that the risk of developing addiction following any smokeless use is lower than for cigarettes.[30,35]

Medicinal nicotine products are intended for short-term use to support quitting smoking and are not designed to be attractive to non-tobacco users or to promote chronic use. The design, dosing and flavour (in the case of oral gum and lozenges) are intended to provide acceptable means of medicinal delivery to motivated smokers. Their nicotine dosing characteristics do not appear to facilitate the development of addiction and, although it is possible in principle, there is no documented evidence that initiation of tobacco addiction through the use of medicinal nicotine products has occurred.

The results of laboratory studies of the abuse liability of nicotine replacement medications are consistent with these conclusions, revealing the medications to be of low abuse liability compared with cigarettes.[36–39] The nicotine nasal spray is the only medicinal form of nicotine replacement that readily enables the user to achieve doses of nicotine more closely approximating those provided by cigarette smoking or oral tobacco use. In fact, the nasal spray is capable of delivering nicotine rapidly enough to produce a transient two-fold nicotine boost in arterial blood relative to venous blood, and although its abuse liability is low compared with cigarettes, it does produce effects beyond those of other marketed medicinal nicotine products.[40,41] These effects led the US Drug Enforcement Administration and the FDA to consider regulating the nasal spray as a controlled substance. Ultimately, the agencies decided against this, in part because the sensory properties of the nasal spray were considered unattractive enough to discourage use or abuse by persons not motivated to quit smoking.[3,31]

Withdrawal

Although separate entities, withdrawal and dependence are closely related in that persons who show signs of dependence are also more likely, if they discontinue smoking, to experience withdrawal symptoms. Furthermore, withdrawal severity is related to the number of cigarettes smoked per day.[1,2] Nonetheless, although withdrawal is associated with dependence, withdrawal is neither necessary nor sufficient for the development of dependence.[17,18] For example, persons maintained on opioids for pain, sedatives for epilepsy and nicotine replacement for tobacco dependence may show signs of withdrawal if the drug is discontinued abruptly, but most such persons do not use the medicine in a manner considered 'out of control' or meeting criteria for 'dependence'.[42] Conversely, some opioid abusers do not use opioids daily, and show little or no withdrawal when abstinent from the drug. Many stimulant dependent people also display little or no withdrawal when not using.[6,18,43]

Withdrawal is generally defined in terms of rebound-like symptoms that are usually the opposite of those produced by administration of the drug. For example, opioid-associated pupillary dilation is replaced by constriction, sedative-associated muscle relaxation by tremors, and nicotine-associated calming by anxiety and irritation.[6,43] Withdrawal syndromes are diagnosed according to specific criteria for each drug class.[17,18] Criteria for tobacco or nicotine withdrawal include dysphoria or depressed mood state, irritability, anger, difficulty concentrating, anxiety and restlessness. Strong and recurrent cravings are also common and are associated with relapse to tobacco use.

The onset of withdrawal symptoms can contribute to the dependence process because resumption of tobacco use quickly relieves the symptoms. For example, a brain imaging study showed that the first three puffs on a cigarette could deliver sufficient nicotine to occupy 70% of $\alpha 4\beta 2$ nicotine receptors in the brain, and that smoking one cigarette resulted in 88% occupation, thus reducing withdrawal-associated cravings and perpetuating the addiction cycle.[44] It is not clear whether slower delivering forms of nicotine, such as uninhaled cigar smoke or oral smokeless tobacco, can saturate brain nicotine receptors so fully or rapidly, and this potential characteristic of inhaled cigarette smoke may help to explain why the cigarette appears to be the most addictive nicotine delivery system. Nicotine replacement medications can reverse withdrawal associated brain function deficits,[13,33,45] but it is unlikely that current dosage formulations provide as rapid or as complete receptor satiation as is achieved by cigarettes.

Most cigarette smokers in Britain and many other developed countries begin smoking in adolescence and graduate to daily use and dependence, smoking between 10 and 30 cigarettes each day by the time they are around 18 years old. Most of these individuals also show signs of withdrawal on abrupt discontinuation of use. In contrast, intermittent smokers (smokers of fewer than five cigarettes per day) and non-daily cigarette smokers may still meet criteria for dependence, but many do not meet criteria for the withdrawal disorder. Withdrawal from smokeless tobacco use is qualitatively similar to that from cigarette smoking, but the magnitude of the symptoms is generally somewhat less than observed in cigarette smokers. This is consistent with the more gradually onsetting and offsetting effects of oral nicotine delivery compared with smoked tobacco.[30,35]

4.4.2 Environmental determinants of tobacco use and addiction

The high risk of addiction that accompanies tobacco use is not a simple consequence of the intrinsic addictiveness of nicotine or even of the cigarette. Marketing and other social influences (for example, product use by social icons and in the media) affect the likelihood of smoking initiation, continuation and quitting, as well as cigarette brand preferences of specific populations (such as targeted minorities, women and youth).[3,46] Social influences and marketing using athletic and health-associated imagery were important factors in reversing the mid-20th century decline in smokeless tobacco use in the United States.[3,47]

The risk of use of any addictive drug, of progression to regular use and addiction, and the probability of cessation, are all influenced by environmental factors.[2,46] Factors important in initiation of drug use include the image of the drug, social forces (particularly the social acceptability of use and perceived prevalence of use in society and among peers), portrayed use in media, access, cost, and the effectiveness of ongoing drug education and prevention programmes.[2] Similarly, cessation can be precipitated and sustained outside treatment (sometimes referred to as 'spontaneous remission'), by social sanctions, access, cost, peers, and legal consequences.[2] For example, the risk of alcohol addiction is heightened in communities with ready access and greater social tolerance for use, and to opioids in anaesthetists and other medical professionals who have easy access to such drugs. Furthermore, environmental factors can strongly influence the prevalence of use and addiction of various forms of a given drug. This is evident in the dramatic shifts in preferred forms of tobacco over the course of the 20th century in England, the United States and Sweden.[20,48] It is also true of other drugs. For example, the relative preference for beer, spirits, red versus white wine, and so on, have shifted in recent decades probably as a function of marketing campaigns and also by reports that moderate drinking of some alcoholic beverages may have health benefits.[49] The opioid analgesic OxyContin® became the preferred form of oxycodone among drug abusers in many countries, in part due to its formulation and in part due to environmental factors including extensive media reporting which implied more addictive effects than were explained by its pharmacology.[50,51] Medicinal nicotine formulations are designed to be less attractive than products used for initiation of nicotine dependence or any use other than to quit smoking and sustain abstinence. However, it is plausible that patterns of abuse could emerge if these products were marketed more aggressively with, for example, messages implying weight control or attention improvement, or with campaigns similar to those used in the past to give cigarettes a highly desirable image through the use of various cultural icons as role models.

Although the risk of addiction is high following any smoking, and is present with all kinds of tobacco use,[52,53] the fact that not all who try smoking escalate to addiction, and that many people who become daily and addicted smokers are able to quit, is encouraging from a public health perspective. In fact, the situation today in which one third to one half of initial users become regular smokers is much more hopeful than in the 1970s, when Michael Russell observed what Jean Cocteau had said of opium in the 1930s, that, 'He who has smoked will smoke', was then true of tobacco smoking.[22] Specifically, relying upon a 1960s survey of UK adults, Russell observed that persons who had smoked only four cigarettes had a 94% chance of smoking for another 30–40 years.[54] This perspective is important because it illustrates the importance of environmental determinants in the natural history of tobacco use. Environmental interventions that reduce the prevalence of tobacco use and the risk of progression to addiction, and stimulate cessation among users, are now increasingly understood and have been discussed in detail elsewhere.[55–58] They include measures such as increasing price, smoke-free policies, tobacco advertising and promotion bans, health warnings and media campaigns, and individual smoking cessation treatment services. Although restricting access to procurement of cigarettes is supported by many tobacco control leaders to reduce use by youth, the practical difficulties in enforcement have lessened but not negated the importance of the seemingly obvious strong strategy.[59] Cessation of smoking by parents appears to support abstinence and motivate cessation by their children.[60] For established cigarette smokers, the marketing of and improved access to treatment for dependence and withdrawal to aid cessation are important in promoting cessation efforts as well as improving the chances of maintaining abstinence.[61–63]

4.4.3 Role of tobacco product design in promoting addiction to smoking

For nicotine, as for other addictive drugs including opioids, cocaine, sedatives and alcohol, the pharmacological effects and liability of developing addiction depend in part on the formulation or vehicle of delivery, which also influences how the drug is used.[64–66] For example, intravenous and smoked preparations of opioids and stimulants produce stronger and more desirable effects (to drug abusers) than oral preparations because the impact is quicker and stronger. Among oral formulations such as tablets, crushable forms that have a faster, stronger effect are generally the most attractive and addictive.[66] Smokable forms of opioids and cocaine, such as crack cocaine, essentially did for these drugs what cigarettes did for tobacco – they made it convenient to repeatedly self-administer desired doses, producing rapid and powerful effects that are extremely addictive.[64,67]

The ingredients and design of tobacco products are intended to facilitate the rapid delivery of nicotine and consequently enhance the development of addiction in several ways.[3–5] Tobacco leaf contains extractable nicotine that can readily lead to addiction by a variety of means of ingestion, though the degradation of nicotine that occurs in the liver following its oral ingestion makes this a relatively unattractive route. Over thousands of years in the Americas, and over centuries throughout the rest of the world, tobacco-utilising vehicles for nicotine delivery have evolved to enhance the addictiveness of the drug to suit many different cultures, needs and interests.[7,68] Furthermore, tobacco companies have researched and developed cigarettes to increase the likelihood that they will induce addiction quickly and strongly.[3–5,9–11] Similarly, smokeless tobacco products have been engineered by many companies to facilitate the probability of repeated use and escalation to addiction. This includes physical design factors to control nicotine as well as to reduce unattractive product features, buffering chemicals to control nicotine dose and speed of absorption, and flavouring to appeal to various target populations (such as cherry flavour for youth).[3,32,47]

A recent report by the Tobacco Regulation Study Group of the World Health Organization concluded that tobacco product modifications including 'candy-like' flavouring have been used to target youth for initiation.[29] Other published reports based on tobacco industry document analysis came to similar conclusions.[26–29] In addition to making cigarettes more attractive to youth through flavour modifica-tion, cigarettes have been designed to provide smoother, less irritant smoke, using various chemical additives and physical design features to ease youth initiation.[26] Menthol has been used to cool and smooth the smoking experience, as well as to provide a distinct sensory feature incorporated in branding efforts targeted at various ethnic or other populations.[69,70] Tobacco companies have targeted women with cigarettes, employing designs and ingredients to make them more physically attractive.[71,72] Moreover, as documented by the US FDA, and through more recent evaluations of tobacco industry documents, the industry conducted extensive and sophisticated research on the neurobiology of tobacco dependence and the importance of nicotine dosing through control of free-base nicotine levels.[3,31,73,74]

The concept of modifying the delivery system to affect addictiveness and attractiveness is neither new nor unique to tobacco. Flavoured alcoholic beverages – so called 'alcopops' – make alcohol more palatable, particularly for young people. Flavouring and other product modifications have been employed in youth-targeted alcohol marketing, according to the National Research Council and Institute of Medicine.[49,75] The drug addiction researcher and editor of *Drug and Alcohol Dependence*, Robert Balster, made the following conclusion in his report to the World Health Organization:

> *It is well known that formulations and routes of administration play significant roles in the abuse of many drugs, including nicotine. Drug abuse scientists have begun to use this knowledge to suggest changes in the formulations of pharma-ceuticals to reduce their risk for abuse and dependence. It is a perversion of*

science that the tobacco industry would apply these same principles to increase the use and abuse of their products.[76]

Tobacco industry documents confirm that the industry has used ammonia compounds to increase the fraction of rapidly absorbable free-base nicotine in smoke; acetaldehyde to interact with nicotine and to produce stronger 'synergistic' effects; menthol, glycerin, leuvenilic acid and chocolate in cigarettes to enable deeper absorption of smoke by masking the noxious constituents including nicotine; and extensive use of flavouring and physical design features to increase attractiveness and facilitate the development of addiction.

4.4.4 Genetic influences on tobacco addiction

Evidence that genetic factors play a role in the vulnerability to developing tobacco addiction as well as difficulty in achieving and sustaining abstinence originates in twin studies in the 1990s.[77–79] These studies support the conclusion that the heritability index, namely the relative contribution of genetic factors or the 'heritability' of cigarette smoking compared with environmental factors, appears to be greater than 50%, with more recent studies suggesting heritability closer to 70%.[80] This is in the range of what is also estimated for alcoholism.[81]

Since then, a number of potential genetic mechanisms have been implicated in the development of tobacco use in youth, in addition to known environmental and family influences, and also in influencing the maintenance of addiction to smoking, including the serotonin transporter promoter polymorphism, various hepatic enzymes, dopamine receptor genotypes and brain-derived neurotrophic factor.[82–88] Several reviews of the contribution of genetic factors to tobacco addition are available, and this rapidly expanding area of research will not be recapitulated in detail in this report.[89,90] Understanding of these effects on the addiction process is, however, still far from clear, and it seems that influences on the development of addiction may be very different from those that affect the maintenance of smoking behaviour in adults.

For example, the Ten Towns Heart Health Study assessed the relation between various measures of cigarette smoking, nicotine intake, genetically determined enzymes that influence nicotine metabolism and the effects of nicotine and dopamine in young smokers.[88] The study found that persons who metabolised nicotine more slowly due to diminished function of the CYP2A6 liver enzyme were more likely to progress to dependence than those who more rapidly metabolised nicotine. This contrasted with results of studies on adult smokers, which report lower dependence and increased cessation among persons with the

CYP2A6 allele.[88,91,92] Also, whereas studies of adults have found that dopaminergic regulating genotypes affect the propensity to nicotine dependence, no such relation was found in young people in the Ten Towns study.[93–95]

There are also some genetically conferred differences in nicotine metabolism and patterns of use across ethnic groups and, potentially, between males and females.[92,96–102] It appears increasingly likely that some polymorphisms alter the vulnerability to tobacco addiction, possibly by modulation of the neurochemical mechanisms of nicotine reinforcement.[90,103] Preliminary research also suggests that there are genetically conferred differences in beneficial response to medications for treating nicotine dependence.[15,104,105] Such research may ultimately lead to the use of genetic testing to match treatment to individuals more effectively, though ethical and practical challenges to implementation will need to be considered and overcome,[106,107] and the practical applicability and cost effectiveness of this approach are still far from proven.[108]

4.5 Developments in imaging and cognitive assessment

In recent years, growth in understanding has come from more extensive evaluation of tobacco and nicotine effects in humans using brain imaging techniques and increasingly sophisticated cognitive assessment. Brain imaging techniques used in diagnostic medicine and for unravelling the pathogenesis of disease and cognitive function have evolved rapidly since the 1970s. The two most frequently used approaches for evaluating tobacco and nicotine in humans are positron emission tomography (PET) and magnetic resonance imaging (MRI), as well as its variant, functional MRI (fMRI). Electroencephalography (EEG) is not referred to as often as imaging, but modern EEG mapping may also be considered along with other approaches that provide a three-dimensional assessment of brain structure and function. These have been reviewed elsewhere.[109–115] This chapter will highlight some recent findings that have implications for understanding addiction as well as potential implications for addressing tobacco addiction through treatment and policy.

4.5.1 Tobacco and nicotine effects on brain structure and function

Among the most important findings of clinical research utilising brain imaging techniques is that the human brain is changed structurally by tobacco exposure, with the change apparently explicable largely by the effects of nicotine. The key structural change is increased density of nicotine receptor binding sites. This has been demonstrated in studies comparing smokers with non-smokers, in post-mortem studies of smokers, and in the direct relationship between amount of smoking and receptor up-regulation.[116–118] Furthermore, another human autopsy study found a direct positive relationship between daily smoking and density of nicotine receptor binding sites in the hippocampus and thalamus.[116] These findings are generally consistent with those reported in animal studies described in Chapter 3.

In addition to changes in brain structure, there are changes in function that appear to mediate reinforcing, cognitive and other effects of nicotine. These include changes in electrical activity identified through electroencephalography, in regional cerebral glucose metabolism and regional blood flow, and release of dopamine in the ventral striatum section of the brain.[113,119–125]

4.5.2 Brain correlates of craving and withdrawal

Although craving is often considered a purely behavioural or psychological effect, it is, in fact, a reliable correlate of physiological withdrawal signs and it can be experimentally elicited and studied.[126–129] Brain imaging studies now make it clear that these responses are physiologically mediated, thus blurring the distinction between so-called physical and psychological effects. Specifically, overnight abstinence-associated craving for cigarettes and cue-elicited cravings are associated with changes in regional cerebral blood flow and metabolism in studies utilising PET methodology.[130–132] Studies employing fMRI techniques have also demonstrated changes in cerebral blood flow associated with cue-elicited craving.[133–137]

4.5.3 Tobacco and nicotine effects on cognitive functioning and attention

It has long been understood that tobacco abstinence among smokers was associated with impaired cognitive function and attention abilities in particular, and that resumption of smoking or nicotine administration could restore performance.[2,138] Imaging studies support the conclusions that the altered brain functioning associated with withdrawal can be reversed by nicotine administration in a variety of forms, and further that, even in non-smokers, the small benefits of increased attention are present and are associated with observable changes in brain function.[13,139–144]

4.5.4 Reversibility of effects: treatment implications

Although acute functional changes such as cue-elicited craving and associated changes in regional cerebral blood flow are reversible, some changes may be long lasting. For example, a study of 67 female cigarette smokers (mean age 25.5 years) found acutely diminished cognitive function and associated electroencephalographic activation and heart rate, with extended evaluation over 31 days revealing little or no recovery.[145] Although animal studies suggest that effects of nicotine on brain structure and function are probably reversible over time, they are of little help at this point in determining the time course of reversal in persons with many years of exposure to nicotine and other smoke constituents. Moreover, insofar as smoke exposure began during the neuroplastic period of adolescence, the concept of reversing to a baseline level may not be appropriate because the brain was changed by nicotine during its own development.

An implication of these findings is that it appears increasingly probable that some smokers may experience very long-term, perhaps lifelong, disruption of brain function, mood and/or cognitive ability following smoking cessation. Such individuals may require similarly long-term treatment support or nicotine maintenance, and this may account for the sustained use of nicotine medications by some ex-smokers, many of whom report that their use is to enable them to maintain abstinence.[146,147] It is also consistent with the conceptualisation of tobacco dependence as a chronic disorder. It is at odds, however, with the strong guidance on the labelling of tobacco dependence treatment products to discontinue use after a certain period (three or six months) and with limitations in duration of reimbursement of product use.

4.6 Insights from new pharmacological developments

Two new medicines and a new category of medicines have been evaluated in human studies of smoking tobacco dependence. The results of these studies, along with the earlier animal studies, contribute further to our understanding of the mechanisms of tobacco addiction.[148,149] Varenicline is a nicotinic acting drug, but its primary actions are on a subset of the family of receptors that are responsive to nicotine. Specifically, it acts as a full agonist on the $\alpha7$ nicotinic receptor, which is involved in nicotine reinforcement as well as cognitive functioning such as memory. It has more limited, partial agonist effects at the $\alpha4\beta7$ nicotinic receptor, which is also involved in reinforcement memory and other cognitive functioning. It is not clear that these receptors are critical in the withdrawal process and clinical data are not yet available to permit full comparison of the efficacy of varenicline with that of medicinal nicotine as withdrawal treatment. The efficacy of varenicline provides additional support for the key role of nicotine in the tobacco dependence process. It also shows that not all actions of nicotine are required for treatment efficacy, implying that not all actions of nicotine are relevant to the dependence process. This concept is not novel, but the data, from preclinical study to human efficacy trials, shed additional light on the receptor level mechanisms of dependence in humans.

The development of the CB1 cannabinoid receptor antagonist rimonabant, and the demonstration in animal models and preliminary human data that it has some effect as a smoking cessation therapy,[150] is a further interesting development. The CB1 receptor mediates the reinforcing and other effects of marijuana through dopamine release. Whereas activation of this receptor by cannabis use is known to increase appetite, blockade of the receptor by rimonabant reduces appetite and supports smoking cessation.[150] These findings are consistent with animal data demonstrating that nicotine potentiates several physiological and behavioural effects of marijuana extract, delta-9-tetrahydrocanabinol, which acts at the CB1 receptor. Whether the CB1 receptor is vital in the production of nicotine dependence or rather is simply another means by which dopamine release is modulated by nicotine is not clear. Medicinal nicotine reduces appetite and body weight as well as urges to smoke.[151] The apparent linkage of tobacco dependence and/or treatment to cannabis dependence also contributes to the understanding that drug dependence disorders and perhaps appetite disorders have common mechanisms that may lead to better common treatments.

4.7 Conclusions

▸ Addiction to nicotine arises from a combination of genetic, environmental and pharmacological factors, but the characteristics of the nicotine delivery system are also crucially important.

▸ Cigarettes are the most addictive tobacco product.

▸ Cigarettes and many other tobacco products have been specifically designed, engineered and marketed to enhance both development and maintenance of addiction.

▸ Medicinal nicotine products are designed and marketed to minimise their addiction potential.

▸ The development of addiction includes changes in brain structure and function that result in cessation-associated withdrawal effects that typically persist for many weeks or longer in some individuals, thereby impairing the ability to achieve and sustain abstinence.

▸ Treatment of dependence and withdrawal can restore brain function, mood, and cognitive abilities, and thereby support cessation, but individuals appear to vary widely in how long they may require treatment, and probably in what forms of treatment are acceptable and effective.

▶ However, some of the changes in brain structure and function in smokers, particularly in those who began smoking when very young, may not be entirely reversible.

▶ Some smokers may never fully overcome their addiction, or even ever be able to quit all nicotine use.

References

1   Royal College of Physicians of London. *Nicotine addiction in Britain.* London: Royal College of Physicians, 2000.

2   US Department of Health and Human Services. *The health consequences of smoking: nicotine addiction.* Report of the Surgeon General. Rockville, MD: US Department of Health and Human Services, Public Health Service, Centers for Disease Control, Center for Health Promotion and Education, Office on Smoking and Health, 1988.

3   Food and Drug Administration. 21 CFR Part 801 *et al.* Regulations restricting the sale and distribution of cigarettes and smokeless tobacco to protect children and adolescents; final rule. *Federal Register* 1996;61:44396–45318.

4   World Health Organization. *Advancing knowledge on regulating tobacco products.* Monograph. Geneva: World Health Organization, 2001.

5   World Health Organization Scientific Advisory Committee on Tobacco Product Regulations. *Recommendation on tobacco product ingredients and emissions.* Geneva: WHO SACTob, 2003.

6   O'Brien CP. Drug addiction and drug abuse. In: Hardman JG, Limbird LE (eds), *Goodman and Gilman's the pharmacological basis of therapeutics*, 10th edn. New York: McGraw-Hill; 2001:621–42.

7   World Health Organization. *Tobacco: deadly in any form or disguise.* World No Tobacco Day 2006. Geneva: World Health Organization, 2006.

8   Russell MA. Cigarette smoking: natural history of a dependence disorder. *Br J Med Psychol* 1971;44:1–16.

9   Slade J, Bero LA, Hanauer P, Barnes DE, Glantz SA. Nicotine and addiction: the Brown and Williamson documents. *J Am Med Assoc* 1995;274:225–33.

10   Hurt RD, Robertson CR. Prying open the door to the tobacco industry's secrets about nicotine: the Minnesota Tobacco Trial. *J Am Med Assoc* 1998;280:1173–81.

11   Bates C, Connolly GN, Jarvis M. Tobacco additives: cigarette engineering and nicotine addiction. *Action on Smoking and Health*, 1999.

12   Langley JN. On the reaction of cells and of nerve-endings to certain poisons, chiefly as regards the reaction of striated muscle to nicotine and to curare. *J Physiol* 1905;83:374–413.

13   Mansvelder HD, van Aerde KI, Couey JJ, Brussaard AB. Nicotinic modulation of neuronal networks: from receptors to cognition. *Psychopharmacology* 2006;184:292–305.

14   Mansvelder HD, McGehee DS. Cellular and synaptic mechanisms of nicotine addiction. *J Neurobiol* 2002;53:606–17.

15   Lerman C, Niaura R. Applying genetic approaches to the treatment of nicotine dependence. *Oncogene* 2002;21:7412–20.

16   Sullivan PF, Kendler KS. The genetic epidemiology of smoking. *Nicotine Tob Res* 1999; (Suppl 2)51–7.

17 World Health Organization. *The ICD-10 classification of mental and behavioural disorders: clinical descriptions and diagnostic guidelines*. Geneva: World Health Organization, 1992.

18 American Psychiatric Association. *Diagnostic and Statistical Manual of Mental Disorders*, 4th edn. 1994.

19 US Department of Health and Human Services. *The health consequences of involuntary smoke exposure*. Report of the Surgeon General. Atlanta, GA: US Department of Health and Human Services, Centers for Disease Control and Prevention, National Center for Chronic Disease Prevention and Health Promotion, Office on Smoking and Health, 2006.

20 US Department of Health and Human Services. *Reducing tobacco use*. Report of the Surgeon General. Atlanta, GA: US Department of Health and Human Services, Centers for Disease Control and Prevention, National Center for Chronic Disease Prevention and Health Promotion, Office on Smoking and Health, 2000.

21 Reboussin BA, Anthony JC. Is there epidemiological evidence to support the idea that a cocaine dependence syndrome emerges soon after onset of cocaine use? *Neuropsychopharmacology* 2006;31:2055–64.

22 McNeill AD. The development of dependence on smoking in children. *Br J Addiction* 1991;86: 589–92.

23 US Department of Health and Human Services. *Preventing tobacco use among young people*. A Report of the Surgeon General. Atlanta, GA: US Department of Health and Human Services, Public Health Service, Centers for Disease Control and Prevention, National Center for Chronic Disease Prevention and Health Promotion, Office on Smoking and Health, 1994.

24 DiFranza JR, Savageau JA, Rigotti NA *et al.* Development of Symptoms of tobacco dependence in youths: 30 month follow up data from the DANDY study. *Tob Control* 2002;11: 228–35.

25 Henningfield JE, Benowitz NL, Connolly GN *et al.* Reducing tobacco addiction through tobacco product regulation. *Tob Control* 2004;13:132–5.

26 Wayne GF, Connolly GN. How cigarette design can affect youth initiation into smoking: Camel cigarettes 1983–93. *Tob Control* 2002;11(Suppl 1):32–9.

27 Carpenter CM, Wayne GF, Pauly JL, Koh HK, Connolly GN. New cigarette brands with flavors that appeal to youth: tobacco marketing strategies. *Health Aff (Millwood)* 2005;24:1601–10.

28 Giovino GA, Yang J, Tworek C *et al. Use of flavored cigarettes among older adolescent and adult smokers: United States, 2004.* Chicago: National Conference on Tobacco or Health, 2005.

29 World Health Organization. Study group on tobacco product regulation (TobReg). *Scientific advisory note on candy flavored tobacco products: research needs and recommended actions by regulators*. Geneva: World Health Organization, 2006.

30 Hatsukami DK, Severson, HH. Oral spit tobacco: addiction, prevention and treatment. *Nicotine Tob Res* 1999;1:21–44.

31 Food and Drug Administration. Regulations restricting the sale and distribution of cigarettes and smokeless tobacco products to protect children and adolescents; proposed rule analysis regarding FDA's jurisdiction over nicotine-containing cigarettes and smokeless tobacco products; notice. *Federal Register* 1995;60:41314–792.

32 Henningfield JE, Radzius A, Cone EJ. Estimation of available nicotine content of six smokeless tobacco products. *Tob Control* 1995;4:57–61.

33 Henningfield JE, Fant RV, Buchhalter AR, Stitzer ML. Pharmacotherapy for nicotine dependence. *CA Cancer J Clin* 2005;55:281–99.

34   Hatsukami DK, Hecht, S *Hope or hazard? What research tells us about 'potentially reduced-exposure' tobacco products.* Minneapolis, MN: University of Minnesota, 2005.

35   Henningfield JE, Fant RV, Tomar SL. Smokeless tobacco: an addicting drug. *Adv Dent Res,* 1997;11:330–5.

36   Henningfield JE, Keenan RM. Nicotine delivery kinetics and abuse liability. *J Consult Clin Psychol* 1993;61:743–50.

37   Houtsmuller EJ, Fant RV, Eissenberg TE, Henningfield JE, Stitzer ML. Flavor improvement does not increase abuse liability of nicotine chewing gum. *Pharmacol Biochem Behav* 2002;72:559–68.

38   Houtsmuller EJ, Henningfield JE, Stitzer ML. Subjective effects of the nicotine lozenge: assessment of abuse liability. *Psychopharmacology* 2003;167:20–7.

39   Stitzer ML, de Wit H. Abuse liability of Nicotine. In: Benowitz NL (ed), *Nicotine safety and toxicity.* New York: Oxford University Press, 1998:1119–31.

40   Gourlay S, Benowitz N. Arteriovenous differences in plasma concentration of nicotine and catecholamines and related cardiovascular effects after smoking, nicotine nasal spray, and intravenous nicotine. *Clin Pharmacol Ther* 1997;62:453–63.

41   Schuh KJ, Schuh LM, Henningfield JE Stitzer ML. Nicotine nasal spray and nicotine vapor inhaler: abuse liability determination. *Psychopharmacology* 1997 130:352–61.

42   Savage SR. Principles of pain treatment in the addicted patient. In: Grahm AW, Schultz TK (eds), *Principles of addiction medicine.* Chevy Chase, Maryland: American Society of Addition Medicine, 1998:919–59.

43   Koob GF, Le Moal M. *Neurobiology of addiction.* London: Academic Press, 1996.

44   Brody AL, Mandelkern MA, London, ED. Cigarette smoking saturates brain {alpha}4beta2 nicotinic acetylcholine receptors. *Arch Gen Psychiatry* 2006;63:907–14.

45   Levin ED, McClernon FJ, Rezvani AH. Nicotinic effects on cognitive function: behavioral characterization, pharmacological specification, and anatomic localization. *Psychopharmacology* 2006;184:523–39.

46   Warner KE, Isaacs SL, Knickman JR. *Tobacco control policy.* Chichester: John Wiley, 2006.

47   Connolly GN. The marketing of nicotine addiction by one oral snuff manufacturer. *Tob Control* 1995;4:73–9.

48   Foulds J, Ramstrom L, Burke M, Fagerstrom K. Effect of smokeless tobacco (snus) on smoking and public health in Sweden. *Tob Control* 2003;12:349–59.

49   National Research Council and Institute of Medicine. *Reducing underage drinking: a collective responsibility.* Committee on developing a strategy to reduce and prevent underage drinking. Bonnie RJ, O'Connell ME (eds). Board on Children, Youth, and Families, Division of Behavioral and Social Sciences and Education. Washington, DC: National Academy Press, 2003.

50   Wright C 4th, Kramer ED, Zalman MA, Smith MY, Haddox JD. Risk identification, risk assessment, and risk management of abusable drug formulations. *Drug Alcohol Depend* 2006;83(Suppl 1):68–76.

51   United States General Accounting Office. *Prescription drugs: OxyContin abuse and diversion and efforts to address the problem.* Report number: GAO-04-110, January 2004. www.gao.gov/htext/d04110.html (accessed 7 August 2007).

52   Wald NJ, Watt HC. Prospective study of effect of switching from cigarettes to pipes or cigars on mortality from three smoking related diseases. *BMJ* 1997;314:1860–3.

53   Boyle P, Ariyaratne M. Curbing tobacco's toll starts with the professionals: World No Tobacco Day. *Lancet* 2005;365:1990–2.

54 Russell MA. The nicotine addiction trap: a 40-year sentence for four cigarettes. *Br J Addiction* 1990;85:293–300.

55 Borland R, Davey C. *Tobacco: science, policy and public health.* Oxford: Oxford University Press, 2004:707–32.

56 Jha P, Ross H, Corrao MA, Chaloupka FJ. *Tobacco: science, policy and public health.* Oxford: Oxford University Press, 2004:733–748.

57 Lantz PM, Jacobson PD, Warner KE *et al.* Investing in youth tobacco control: a review of smoking prevention and control strategies. *Tob Control* 2000;9:47–63.

58 Warner KE, Jacobson PD, Kaufman NJ. Innovative approaches to youth tobacco control: introduction and overview. *Tob Control* 2003;12(Suppl 1):1–5.

59 Wakefield M, Giovino G. Ten penalties for tobacco possession, use, and purchase: evidence and issues. *Tob Control* 2003;12(Suppl 1):6–13.

60 Henningfield JE, Moolchan ET, Zeller M. Regulatory strategies to reduce tobacco addiction in youth. *Tob Control* 2003;12:14–24.

61 The World Bank. *Development in practice: curbing the epidemic. Governments and the economics of tobacco control.* Washington, DC: The World Bank, the International Bank for Reconstruction and Development, 1999.

62 US Centers for Disease Control and Prevention. Impact of promotion of the great american smokeout and availability of over-the-counter nicotine medications. *MMWR Weekly* 1997;46:867–71.

63 World Health Organization. *Policy recommendations for smoking cessation and treatment of tobacco dependence.* Geneva: World Health Organization, 2004.

64 Hardman JG, Limbird LE (eds). *Goodman and Gilman's the pharmacological basis of therapeutics,* 9th edn. New York: McGraw-Hill, 1996.

65 Compton WM, Volkow ND. Abuse of prescription drugs and the risk of addiction. *Drug Alcohol Depend* 2006;83(Suppl 1):4–7.

66 Grudzinskas C, Balster RL, Gorodetsky CW *et al.* Impact of formulation on the abuse liability, safety and regulation of medications: the expert panel report. *Drug Alcohol Depend* 2006; 83(Suppl 1):77–82.

67 Cone EJ. Pharmacokinetics and pharmacodynamics of cocaine. *J Analytical Toxicol* 1995; 19:459–78.

68 Wilbert J. *Tobacco and Shamanism in South America.* New Haven: Yale University Press, 1987.

69 Wayne GF, Connolly GN. Application, function, and effects of menthol in cigarettes: A survey of tobacco industry documents. *Nicotine Tob Res* 2004;6:S43–54.

70 Henningfield JE, Benowitz NL, Ahijevych K *et al.* Does menthol enhance the addictiveness of cigarettes? An agenda for research. *Nicotine Tob Res* 2003;5:9–11.

71 Carpenter CM, Wayne GF, Connolly GN. Designing cigarettes for women: new findings from the tobacco industry documents. *Addiction* 2005;100:837–51.

72 Henningfield JE, Santora PB, Stillman FA. Exploitation by design – could tobacco industry documents guide more effective smoking prevention and cessation for women? *Addiction* 2005; 100:735–6.

73 Wayne GF, Connolly GN, Henningfield JE. Assessing internal tobacco industry knowledge of the neurobiology of tobacco dependence. *Nicotine Tob Res* 2004;6:927–40.

74 Wayne GF, Connolly GN, Henningfield JE. Brand differences of free-base nicotine delivery in cigarette smoke: the view of the tobacco industry documents. *Tob Control* 2006;15:189–98.

75   World Health Organization. *Global status report on alcohol*. Geneva: World Health
      Organization, 1999.

76   Balster RL. *Commentary on the relation of cigarette design to addictiveness and toxicity of
      cigarette products: relevance of formulation to drug abuse liability from the perspective of
      nonnicotine pharmaceuticals and drugs of abuse*. Meeting of the Study Group on Tobacco
      Product Regulation, June 2006.

77   Carmelli D, Swan GE, Robinette D, Fabsitz R. Genetic influence on smoking – a study of
      male twins. *New Engl J Med* 1992;327:829–33.

78   Heath AC, Martin NG. Genetic models for the natural history of smoking: evidence for a
      genetic influence on smoking persistence. *Addictive Behav* 1993;18:19–34.

79   Kendler KS, Neale MC, Sullivan P *et al*. A population-based twin study in women of
      smoking initiation and nicotine dependence. *Psychol Med* 1999:299–308.

80   Vink JM, Willemsen G, Boomsma DI. Heritability of smoking initiation and nicotine
      dependence. *Behav Genet* 2005;35:397–406.

81   Schuckit MA. Genetics of the risk for alcoholism. *Am J Addictions* 2000;9:103–112.

82   Rodriguez S, Huang S, Chen X *et al*. A study of TH01 and IGF2-INS-TH haplotypes in
      relation to smoking initiation in three independent surveys. *Pharmacogenet Genomics*
      2006;16:15–23.

83   Timberlake DS, Haberstick BC, Lessem JM. An association between the DAT1
      polymorphism and smoking behavior in young adults from the national longitudinal
      study of adolescent health. *Health Psychology* 2006;25:190–7.

84   Rende R, Slomkowski C, McCaffery J, Lloyd-Richardson EE, Niaura R. A twin-sibling
      study of tobacco use in adolescence: etiology of individual differences and extreme scores.
      Nicotine Tob Res 2005;7:413–9.

85   Pergadia ML, Heath AC, Agrawal A *et al*. The implications of simultaneous smoking
      initiation for inferences about the genetics of smoking behavior from twin data. *Behav
      Genet* 2006;36:567–76.

86   Clayton RR, Merikengas KR, Abrams DB. Editorial: introduction to tobacco, nicotine and
      youth: the Tobacco Etiology Research Network. *Drug Alcohol Depend* 2000;59(Suppl 1):1–4.

87   Zhang L, Kendler KS, Chen X. Association of the phosphotase and tensin homolog gene
      (PTEN) with smoking initiation and nicotine dependence. *Am J Med Genet B
      Neuropsychiatr Genet* 2006;141:10–14.

88   Huang S, Cook DG, Hinks LJ *et al*. CYP2A6, MAO4, DBH, DRD4, 5HT2A genotypes,
      smoking behaviour and cotinine levels in 1518 UK adolescents. *Pharmacogenet Genomics*
      2005; 15:839–50.

89   Sullivan PF, Kendler KS. The genetic epidemiology of smoking. *Nicotine Tob Res*
      1999;1(Suppl 2):51–7.

90   Gerra G, Garafano L, Zaimovic A *et al*. Association of the serotonin transporter promoter
      polymorphism with smoking behavior among adolescents. *Am J Med Genet B
      Neuropsychiatr Genet* 2005;135:73–8.

91   Tyndale RF, Sellers EM. Variable CYP2A6-mediated nicotine metabolism alters smoking
      behavior and risk. *Drug Metab Dispos* 2001;29:548–52.

92   Schoedel KA, Hoffman EB, Rao Y, Sellers EM, Tyndale RF. Ethnic variation in CYP2A6
      and association of genetically slow nicotine metabolism and smoking in adult Caucasians.
      Pharmacogenet 2004:14:615–26.

93   McKinney EF, Walton RT, Yudkin P. Association between polymorphisms in dopamine
      metabolic enzymes and tobacco consumption in smokers. *Pharmacogenet* 2000;10:
      483–91.

94 Ito H, Hamajima N, Matsuo K *et al.* Monoamine oxidase polymorphisms and smoking behaviour in Japanese. *Pharmacogenet* 2003;13:73–9.

95 Shields PG, Lerman C, Audrain J, Bowman ED, Main D, Boyd NR, Caporaso NE. Dopamine D4 receptors and the risk of cigarette smoking in African-Americans and Caucasians. *Cancer Epidemiol Biomarkers Prev* 1998;7:453–8.

96 Kendler KS, Thornton LM, Pedersen NL. Tobacco consumption in Swedish twins reared apart and reared together. *Archs Gen Psychiatry* 2000;57:886–92.

97 Perkins KA, Donny E, Caggiula AR. Sex differences in nicotine effects and self-administration: review of human and animal evidence. *Nicotine Tob Res* 1999;1:301–15.

98 Jin Y, Chen D, Hu Y *et al.* Association between monoamine oxidase gene polymorphisms and smoking behaviour in Chinese males. *Int J Neuropsychopharmacol* 2006;9:557–94.

99 Ahijevych KL, Tyndale RF, Dhatt RK, Weed HG, Browning KK. Factors influencing cotinine half-life during smoking abstinence in African American and Caucasian women. *Nicotine Tob Res* 2002;4:423–31.

100 Beuten J, Ma JZ, Payne TJ *et al.* Significant association of BDNF haplotypes in European-American male smokers but not in European-American female or African-American smokers. *Am J Med Genet B Neuropsychiatr Genet* 2005;139:73–80.

101 Mwenifumbo JC, Myers MG, Wall TL *et al.* Ethnic variation in CYP2A6*7, CYP2A6*8 and CYP2A6*10 as assessed with a novel haplotyping method. *Pharmacogenet Genomics* 2005:15:189–92.

102 Gaedigk A, Casley WL, Tyndale RF *et al.* Cytochrome P4502C9 (CYP2C9) allele frequencies in Canadian Native Indian and Inuit populations. *Can J Physiol Pharmacol* 2001;79:841–7.

103 Howard LA, Ahluwalia JS, Lin SK, Sellers EM, Tyndale RF. CYP2E1*1D regulatory polymorphism:association with alcohol and nicotine dependence. *Pharmacogenetics* 2003;13:321–8.

104 Lerman C, Jepson C, Wiley EP. Role of functional genetic variation in the dopamine D2 receptor (DRD2) in response to bupropion and nicotine replacement therapy for tobacco dependence: results of two randomized clinical trials. *Neuropsychopharmacol* 2006;31:231–42.

105 Swan GE, Valdes AM, Ring HZ *et al.* Dopamine receptor DRD2 genotype and smoking cessation outcome following treatment with bupropion SR. *Pharmacogenomics J* 2005;5:21–9.

106 Caron L, Karkazis K, Raffin TA, Swan G, Koenig BA. Nicotine addiction through a neurogenomic prism: ethics, public health and smoking. *Nicotine Tob Res* 2005;7:181–97.

107 Shields AE, Blumenthal D, Weiss KB *et al.* Barriers to translating emerging genetic research on smoking into clinical practice. Perspectives of primary care physicians. *J Gen Inter Med* 2005;20:131–8.

108 Berrettini WH, Lerman CE. Pharmacotherapy and pharmacogenetics of nicotine dependence. *Am J Psychiatry* 2005;162:1441–51.

109 London ED. *Mapping the cerebral metabolic responses to nicotine. Brain imaging of nicotine and tobacco smoking.* Ann Arbor: NPP Books, 1995:153–66.

110 Volkow ND, Fowler JS, Ding YS, Wang, GJ, Gatley SJ. Imaging the neurochemistry of nicotine actions: studies with positron emission tomography. *Nicotine Tob Res* 1999;1:S127–32.

111 McClernon FJ, Gilbert DG. Human functional neuroimaging in nicotine and tobacco research: basics, background and beyond. *Nicotine Tob Res* 2004;6:941–59.

112 Gatley SJ, Volkow ND, Wang G *et al.* PET imaging in clinical drug abuse research. *Curr Pharm Des* 2005;11:3203–19.

113 Domino EF. Effects of tobacco smoking on electroencephalographic, auditory evoked and event related potentials. *Brain Cognition* 2003;53:66–74.

114 Honey G, Bullmore E. Human pharmacological MRI. *Trends Pharmacol Sci* 2004;25:366–74.

115 Frankle WG, Slifstein M, Talbot PS, Laruelle M. Neuroreceptor imaging in psychiatry: theory and applications. *Int Rev Neurobiol* 2005;67:385–440.

116 Breese CR, Marks MJ, Logel J *et al.* Effect of smoking history on [3H]nicotine binding in human postmortem brain. *J Pharmacol Exp Ther* 1997;282:7–13.

117 Brody AL, Mandelkern MA, Jarvik ME *et al.* Differences between smokers and nonsmokers in regional gray matter volumes and densities. *Biol Psychiatry* 2004:55:77–84.

118 Perry DC, Davila-Garcia MI, Stockmeier CA, Kellar KJ. Increased nicotinic receptors in brains from smokers: membrane binding and autoradiography studies. *J Pharmacol Exp Ther* 1999:289:1545–52.

119 Knott VJ. Electroencephalographic characterization of cigarette smoking behavior. *Alcohol* 2001;24:95–7.

120 Martin-Solch C, Magyar S, Kunig G. Changes in brain activation associated with reward processing in smokers and nonsmokers. A positron emission tomography study. *Exp Brain Res* 2001;139:278–86.

121 Stapleton JM, Gilson SF, Wong DF *et al.* Intravenous nicotine reduces celebral glucose metabolism: a preliminary study. *Neuropsychopharmacol* 2003;28:765–72.

122 Rose JE, Behm FM, Westman EC *et al.* PET Studies of the influences of nicotine on neural systems in cigarette smokers. *Am J Psychiatry* 2003;160:323–33.

123 Barrett SP, Boileau I, Okker J, Pihl RO, Dagher A. The hedonic response to cigarette smoking is proportional to dopamine release in the human striatum as measured by positron emission tomography and [11C] raclopride. *Synapse* 2004;54:65–71.

124 Brody AL, Olmstead RE, London ED *et al.* Smoking-induced ventral striatum dopamine release. *Am J Psychiatry* 2004;161:1211–8.

125 Zubieta JK, Heitzeg MM, Xu Y *et al.* Regional cerebral blood flow responses to smoking in tobacco smokers after overnight abstinence. *Am J Psychiatry* 2005;162:567–77.

126 Carter Bl, Tiffany ST. Cue-reactivity and the future of addiction research. *Addiction* 1999 Mar;94:349–51.

127 Carter BL, Tiffany ST. The cure-availability paradigm: the effects of cigarette availability on cue reactivity in smokers. *Exp Clin Psychopharmacol.* 2001:9:183–90.

128 Shiffman S. Comments on craving. *Addiction* 2000;95(Suppl 2):171–5.

129 Shiffman S, Engberg JB, Paty JA *et al.* A day at a time: predicting smoking relapse from daily urge. *J Abnorm Psychol* 1997;106:104–16.

130 Brody AL, Mandelkern MA, London ED *et al.* Brain metabolic changes during cigarette craving. *Arch Gen Psychiatr* 2002;59:1162–72.

131 Brody AL, Mandelkern MA, Lee G *et al.* Attenuation of cue-induced cigarette craving and anterior cingulate cortex activation in bupropion-treated smokers: a preliminary study. *Psychiatry Res* 2004;130:269–81.

132 Zubieta JK, Heitzeg MM, Xu Y *et al.* Regional cerebral blood flow responses to smoking in tobacco smokers after overnight abstinence. *Am J Psychiatry* 2005:162:567–77.

133 Lee J, Lim Y, Wiederhold BK, Graham SJ. A functional magnetic resonance imaging (fMRI) study of cue-induced smoking craving in virtual environments. *Appl Psychophysiol Biofeedback* 2005;30:195–204.

134 Wilson SJ, Sayette MA, Delgado MR, Fiez JA. Instructed smoking expectancy modulates cue-elicited neural activity: a preliminary study. *Nicotine Tob Res* 2005;7:637–45.

135 Smolka MN, Buhler M, Klein S *et al.* Severity of nicotine dependence modulates cue-induced brain activity in regions involved in motor preparation and imagery. *Psychopharmacol* 2006;184:577–88.

136 Due DL, Huettel SA, Hall WG, Rubin DC. Activation in mesolimbic and visuospatial neural circuits elicited by smoking cues: evidence from functional magnetic resonance imaging. *Am J Psychiatry* 2002;159:954–60.

137 Shinohara T, Nagata K, Yokoyama E *et al.* Acute effects of cigarette smoking on global cerebral blood flow in overnight abstinent tobacco smokers. *Nicotine Tob Res* 2006;8:113–21.

138 Heishman SJ, Taylor RC, Henningfield JE. Nicotine and smoking: a review of effects on human performance. *Exp Clin Psychopharmacol* 1994;2:345–95.

139 Knott VJ, Harr A. Assessing the topographic EEG changes associated with aging and acute/long-term effects of smoking. *Neuropsychobiology* 1996;33:210–22.

140 Kumari V, Gray JA, Ffytche DH *et al.* Cognitive effects of nicotine in humans: a fMRI study. *Neuroimage* 2003;19:1002–13.

141 Lawrence NS, Ross TJ, Stein EA. Cognitive mechanisms of nicotine on visual attention. *Neuron* 2002;36:539–48.

142 Levin ED, McClernon FJ, Rezvani AH. Nicotinic effects on cognitive function: behavioral characterization, pharmacological specification, and anatomic localization. *Psychopharmacology* 2006;184:523–39.

143 Thiel CM, Zilles K, Fink GR. Nicotine modulates reorienting of visuospatial attention and neural activity in human parietal cortex. *Neuropsychopharmacology* 2005;30:810–20.

144 Xu J, Mendrek A, Cohen MS *et al.* Brain activity in cigarette smokers performing a working memory task: effect of smoking abstinence. *Biol Psychiatry* 2005;58:143–50.

145 Gilbert D, McClernon J, Rabinovich N *et al.* Effects of quitting smoking on EEG activation and attention last for more than 31 days and are more severe with stress, dependence, DRD2 A1 allele, and depressive traits. *Nicotine Tob Res* 2004;6:249–67.

146 Shiffman S, Hughes JR, Pillitteri JL, Burton SL. Persistent use of nicotine replacement therapy: an analysis of actual purchase patterns in a population based sample. *Tob Control* 2003; 12:310–6.

147 Shiffman S, Hughes JR, Di Marino ME, Sweeney CT. Patterns of over-the-counter nicotine gum use: persistent use and concurrent smoking. *Addiction* 2003;98:1747–53.

148 Fagerstrom K, Balfour DJ. Neuropharmacology and potential efficacy of new treatments for tobacco dependence. *Expert Opin Investig Drugs* 2006;15:107–16.

149 Foulds J, Steinberg MB, Williams JM, Ziedonis DM. Developments in pharmacotherapy for tobacco dependence: past, present and future. *Drug Alcohol Rev* 2006;25:59–71.

150 Gelfand EV, Cannon CP. Rimonabant: a cannabinoid receptor type 1 blocker for management of multiple cardiometabolic risk factors. *J Am Coll Cardiol* 2006;47:1919–26.

151 Fiore MC, Bailey WC, Cohen SJ *et al. Treating tobacco use and dependence.* Clinical Practice Guideline. Rockville, MD: US Department of Health and Human Services. Public Health Service, June 2000.

5 | Sources of nicotine for human use

5.1   Available sources of nicotine for human use
5.2   Dose and delivery kinetics of nicotine from different sources
5.3   Contaminants and additives
5.4   Addiction potential of alternative nicotine products
5.5   Use of alternative nicotine sources as substitutes for cigarettes
5.6   Conclusions

5.1 Available sources of nicotine for human use

Nicotine is available to people in the form of conventional tobacco products, nicotine medications and, more recently, in cigarette-like smoking devices that are intended to reduce exposure to some tobacco toxins, also called potential reduced exposure products (PREPs).

5.1.1 Tobacco

Tobacco generally refers to the leaves and other parts of plants that have been domesticated and used to obtain the alkaloid nicotine. Tobacco plants are a species of the genus *Nicotiana*, belonging to the Solanaceae (nightshade) family. There are 64 *Nicotiana* species; the two that are cultivated for tobacco are *Nicotiana tobaccum* and *Nicotiana rustica*, the latter containing higher levels of nicotine.

Nicotiana tobaccum is the major source of commercial tobacco. Tobacco is consumed in a variety of forms, including cigarettes, cigars, cigarillos, bidis, kreteks, pipe tobacco, snuff (oral and nasal) and chewing tobacco, but the great majority of tobacco consumed around the world is in the form of cigarettes (Fig 5.1).

Cigarettes are composed of shredded tobacco leaves and other parts of tobacco plants, with flavourings and other ingredients added, encased in paper and often attached to a filter through which the smoke is puffed. Most cigarettes are made of blond tobacco, while some European and Middle Eastern cigarettes are made of dark tobacco. Cigars contain dark tobacco leaves that are rolled and coupled with a wrapper that may be tobacco leaf or reconstituted tobacco sheet. Cigarillos are small, narrow cigars with no cigarette paper or filter. Pipe tobacco consists of a

(a) Cigarettes

(e) Oral moist tobacco (snus)

(b) Cigar

(f) Oral compressed tobacco

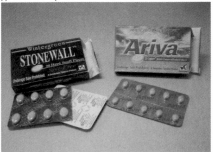

(c) Nasal snuff

(g) Bidis

(d) Oral loose tobacco

(h) Alternative smoking devices (PREPs)

Fig 5.1 Examples of tobacco products.

blend of a variety of leaf types and is often heavily flavoured to give particular aromas and tastes. Bidis (used particularly in India and South East Asia) and kreteks (Indonesia) are described in Chapter 1.

Smokeless tobacco includes oral snuff, loose leaf tobacco, plug tobacco, twist or rope chewing tobacco and dry snuff. Loose leaf tobacco consists of tobacco leaf that has been heavily treated with liquorice and sugars. Plug tobacco is produced from leaves that are immersed in a mixture of liquorice and sugar and then pressed into a plug (usually a square). Twist or rope tobacco is made from leaves flavoured and twisted to resemble a rope.

Moist snuff consists of fine particles of tobacco that contain considerable moisture. Many are treated with flavouring such as wintergreen or mint. Moist snuff is used orally in a technique known as 'snuff dipping', which involves placing a pinch of tobacco between the cheek or lips and the gum, or beneath the tongue. Moist snuff is also available prepared in small porous packets containing a single dose, for use in the same way. Dry snuff is powdered tobacco that usually contains flavour and aroma additives, and is taken by sniffing into the nose. Recently, oral tobacco preparations consisting of tablets of ground and compressed tobacco for buccal administration have been developed (Fig 5.1).

5.1.2 Nicotine medications

Nicotine medications have been developed for use as an aid to smoking cessation. They consist of nicotine chewing gum, nicotine patches, nicotine inhaler, nicotine spray and nicotine lozenges. Nicotine Polacrilex (nicotine gum) was the first of these products to be developed and has been available in a range of doses for well over 20 years. Nicotine transdermal patches became available in the early 1990s and deliver nicotine through the skin over 16- or 24-hour periods. The nicotine inhalator is designed to resemble a cigarette, and delivers nicotine from a cartridge by sucking through a plastic mouthpiece. The nicotine nasal spray is an aqueous solution of nicotine, buffered to be alkaline to facilitate rapid absorption when sprayed into the nose. Nicotine lozenges are small tablets that are placed in the buccal cavity, from which nicotine is absorbed slowly (Fig 5.2).

5.1.3 Potential reduced exposure products

Two tobacco companies in the United States have test-marketed cigarette-like products that deliver nicotine but are reported to deliver less tar and some other tobacco toxins. RJ Reynolds introduced the 'Premier' device, which heats tobacco and tobacco flavour beads using a charcoal ignition system. This results in vapor-

(a) Gum

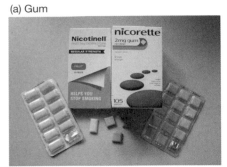

(d) Lozenge

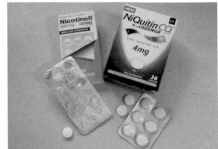

(b) Transdermal patch

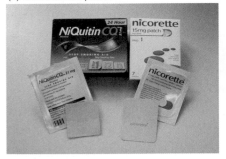

(e) Microtab

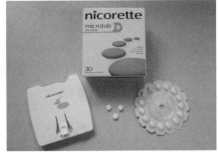

(c) Nasal spray

(f) Inhaler

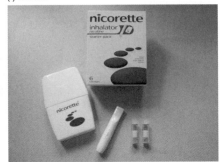

Fig 5.2 Medicinal nicotine products (UK branded examples).

isation of liquid and then condensation into an aerosol. A second generation of this product, 'Eclipse', uses a carbon tip to heat a mixture of tobacco and glycerin which then passes through a charcoal filter.[1] Eclipse differs from Premier in that it does not have flavour beads or an aluminium cylinder. These changes modify the smoke aerosol so that it more closely resembles that of traditional cigarette products.

Philip Morris has test-marketed a device called 'Accord'. This comprises a heating device called a 'puff-activated lighter' which fits special Accord cigarettes. A puff on the cigarette activates a microchip which produces a controlled two-second burn and delivers smoke to the user.[2]

Some cigarettes have been marketed as PREPs based on lower levels of some tobacco smoke carcinogens (Omni and Advance were marketed in the US as low nitrosamine cigarettes),[3,4] and some smokeless tobacco products are marketed as PREPs based on low nitrosamine levels (such as Ariva, a compressed tobacco tablet) (Fig 5.1).[5]

5.2 Dose and delivery kinetics of nicotine from different sources

In considering the dose of nicotine from tobacco and other products, it is important to understand the difference between the nicotine content of the product and the systemic dose of nicotine delivery to the user. The nicotine content refers to how much nicotine is contained in the product either by weight of tobacco or in a unit of use (such as a cigarette). Delivery, or systemic dose, refers to the absolute amount of nicotine absorbed into the body of the user. The term 'bioavailability' will be used to indicate the fraction of the nicotine content of a single dosing unit that is absorbed into the circulation.

5.2.1 Cigarettes

Most cigarettes used by smokers are commercially manufactured products, but there is also considerable use of hand-rolled cigarettes. Most of the data on nicotine delivery come from manufactured cigarettes. Manufactured cigarettes typically contain about 0.7 g of tobacco, but this varies widely according the size of the cigarette and the nature of the processing (whether expanded or 'puffed' tobacco is used). There is wide variation in the nicotine content of the tobacco in cigarettes, ranging from 7 mg/g to 23 mg/g in tobacco sampled from cigarettes from several countries.[6,7] Typical US and UK cigarettes contain about 15–18 mg nicotine per gram of tobacco, or an average of 10–12 mg nicotine per cigarette.

The amount of nicotine absorbed from the cigarette depends on the intensity and volume of puffing, how much smoke is inhaled, and how many puffs are taken from the cigarette. The amount of smoke taken in may also be greatly influenced by whether ventilation holes in the filter are blocked by fingers or lips. Typically, between 1 mg and 2 mg of nicotine enters the systemic circulation from a cigarette, but the delivery of different products varies between 0.5 mg and over 3 mg of nicotine per cigarette.[8]

Cigarettes are routinely tested for 'yield' using the International Organization for Standardization (ISO) or US Federal Trade Commission (FTC) test procedures. These procedures involve the use of machines to smoke cigarettes according to a standard protocol, drawing 35 ml of smoke over two seconds every 60 seconds, until the cigarette is smoked to a 23 mm butt length. Machine test yields have been the basis for marketing of light or mild (low-yield) cigarettes. These cigarettes are engineered to produce low levels of nicotine and tar when tested in smoking machines by ensuring that the smoke drawn into the machine is diluted by ambient air. This is achieved by using more porous cigarette paper and/or introducing ventilation holes in the filter (Fig 5.3), by increasing the burn rate of the cigarette, so the machine takes fewer puffs, and by the use of expanded tobacco.[9] The nicotine content of the tobacco in the cigarette, however, is similar in low yield and regular cigarettes. Thus, the smoker can easily obtain the same dose of nicotine from low yield as from regular cigarettes by taking bigger and more frequent puffs, or by blocking ventilation holes in the filter with fingers or lips.

Fig 5.3 Ventilation holes in the cigarette filter reduce machine-measured nicotine and tar yield.

When a smoker puffs a cigarette, combustion is incomplete, and the smoke produced contains a mixture of tar, water and gases. Particulates form when nicotine vaporises, cools and condenses along with other combustion products. Tobacco smoke is inhaled into the lung, where nicotine is absorbed rapidly into the pulmonary circulation and hence directly into the arterial supply, reaching the brain within a few seconds. As a result of the inhaled route of delivery, nicotine concentrations in arterial blood after inhaling from a cigarette are several times higher than in venous blood, and much higher than those achieved by other nicotine delivery devices.[10,11] This rapid absorption enhances the addictiveness of the delivery system.

5.2.2 Cigars

Cigars vary considerably in size and weight. Typical cigars weigh between 5 and 17 g, but small cigars can weigh as little as 0.5 g and the largest as much as 22 g.[12,13]

The nicotine content of cigars shows similar variation, from around 5 mg to more than 400 mg per cigar. There is also variability in the pH of cigar smoke: small cigars tend to produce acidic pH similar to cigarettes whereas large cigar smoke is more alkaline. Alkaline smoke facilitates absorption of nicotine through the buccal mucosa, while substantial absorption of acidic smoke requires inhalation into the lungs.

Tests of cigars using standard smoking machine parameters defined by the International Committee for Cigar Smoke Study (20 ml puffs taken over 1.5 seconds once every minute until the cigar is burned to a specified butt-length) show that the yields of nicotine and tar for regular cigars are several-fold higher than cigarettes. Nicotine yields average 3.8 mg, 9.8 mg and 13.3 mg for small cigars, cigars and premium cigars, respectively.[12] However, the delivery of nicotine to the smoker depends strongly on whether the smoke is inhaled. Many cigar smokers, particularly those who have never smoked cigarettes, hold the smoke in the mouth without further inhalation, in which case the amount of nicotine absorbed may be relatively small.[14] On the other hand, smokers of small cigars, and particularly those who have been regular cigarette smokers, often inhale the smoke, resulting in high levels of nicotine and smoke toxin exposure.

5.2.3 Pipes

A typical dose of tobacco to fill a pipe bowl is 3–4 g. Pipe smokers and cigarette smokers have similar levels of serum cotinine (the principal metabolite of nicotine) and urine nicotine, indicating that daily nicotine intake is similar for the two forms of tobacco use.[15]

Waterpipes, also known as hookah, narghile, shisha and hubble bubble, are widely used to smoke tobacco in the Middle East and parts of Asia, and increasingly in recent years in Western nations.[16] Tobacco, often flavoured with fruit, mint or cappuccino, is burnt by charcoal and the smoke is drawn through water and then inhaled via a hose with a mouthpiece. An average amount of tobacco per pipe smoked is 20 g. Plasma nicotine levels after smoking for 45 minutes are reported to be higher than those seen after smoking a cigarette.[17] Urinary cotinine levels in regular waterpipe users are similar to those of smokers.[18,19]

5.2.4 Bidis

Typically, each bidi cigarette contains about 0.2 g of tobacco. The nicotine content of the tobacco ranges from 10 mg/g to 17 mg/g and differs in different countries.[6,20] Yields by machine testing of Indian bidis range from 1.9 mg to 2.8 mg nicotine, and

26 mg to 41 mg tar per bidi. Plasma nicotine concentration in people smoking bidis is similar to those seen in people smoking regular cigarettes.[20]

5.2.5 Kreteks

Kreteks are small cigars containing about 60% tobacco, 40% cloves and cocoa. Little data are available on nicotine content and nicotine blood concentrations while smoking kreteks. Machine yields for kreteks range from 2.2 mg to 4.5 mg nicotine and from 48 mg to 113 mg tar per kretek.[21]

5.2.6 Smokeless tobacco

Smokeless tobacco products vary considerably in nicotine content, pH and levels of various carcinogens. The highest concentrations of nicotine are in dry snuff (average 16.8 mg/g), followed by moist snuff (12.6 mg/g) and chewing tobacco (9.9 mg/g).[22] The pH of smokeless tobacco products also varies widely, which is important because pH determines the proportion of nicotine that is in the un-ionized or free-base form, which in turn determines the ease with which nicotine crosses the buccal mucosa and is absorbed into the circulation. In one study of six popular brands of moist snuff from the US, nicotine content ranged from 3.4 mg/g to 11.5 mg/g, while pH ranged from 5.24 to 8.35.[23]

A typical dose of moist tobacco is about 2 g.[24] In one study, the systemic dose of nicotine delivered by 2.5 g of moist snuff held in the mouth for 30 minutes averaged 3.6 mg, and the dose from an average of 7.9 g of chewing tobacco (a typical self-selected dose) chewed for 30 minutes averaged 4.5 mg nicotine.[25] With smokeless tobacco use for 30 minutes, plasma nicotine concentrations rise progressively over the period of use, followed by a slow decline over the next 90 minutes.[25,26] There is evidence that nicotine continues to be absorbed through either the buccal mucosa or from the gastrointestinal tract after the product is removed from the mouth.[25] Studies measuring urine cotinine indicate that regular users of Swedish snus (oral moist snuff) absorb similar total daily doses of nicotine to regular cigarette smokers.[27]

In addition to the traditional smokeless tobacco products, compressed tobacco tablets, such as Ariva and Stonewall, have recently been marketed as alternative oral nicotine products. Ariva contains 2.4 mg nicotine per gram, or approximately 0.7 mg per tablet.[28]

5.2.7 Nicotine medications

Nicotine Polacrilex gum contains 2 mg or 4 mg nicotine. The systemic bioavailability of nicotine from gum averages about 50%, such that the systemic dose is about 1 mg or 2 mg from the 2 mg and 4 mg gum respectively.[29] Systemic bioavailability is less than 100% because not all of the nicotine is extracted from the gum and because a considerable fraction of the nicotine is swallowed. Much of the nicotine absorbed after swallowing is metabolised during the first pass through the liver.

Nicotine patches contain different quantities of nicotine depending on their design. While the nominal dose of nicotine from patches is typically 15 mg for 16 hours or 21 mg for 24 hours, there are differences in actual plasma nicotine concentrations achieved from different brands of patches with the same nominal yields. The absorption of nicotine from patches is gradual with peak concentration times varying from one manufactured patch to another.[30] The rate and extent of the absorption of nicotine from patches may be affected by ambient temperature and by skin blood flow; for example, the rate of absorption may be accelerated by exercise.

Nicotine nasal sprays contain 0.5 mg nicotine per 0.5 ml spray. The recommended dose is one spray in each nostril (total of 1 mg). The bioavailability of nicotine nasal spray averages about 50%, but varies considerably from person to person. Nicotine is absorbed into the systemic circulation faster from nicotine nasal spray than from any other currently available medicinal nicotine preparations (see Chapter 7).[11]

A nicotine inhalator cartridge contains 10 mg nicotine. Each puff delivers about 13 µg of nicotine, which is comparable to approximately one tenth of the dose of nicotine from a puff of a cigarette. The average systemic dose of nicotine delivered by each cartridge is 2 mg.[31]

Nicotine lozenges contain 2 mg or 4 mg nicotine. The bioavailability is slightly greater than that of nicotine gum but the time course of absorption is similar to that of the gum.[32]

5.2.8 Potential reduced exposure products

The Eclipse device, when tested under FTC smoking conditions, generates a yield of 0.18 mg nicotine and 3.2 mg tar. Studies of individuals smoking Eclipse show similar plasma nicotine concentrations to those smoking the usual brand of cigarettes, although carbon monoxide levels are higher than in those using the usual brand.[33] The Accord device yields 0.2 mg nicotine and 3 mg tar by standardised machine testing. Smokers who switch to Accord have lower blood nicotine concentrations than when smoking cigarettes.[34]

5.3 Contaminants and additives

5.3.1 Additives and tobacco

Besides tobacco, numerous other ingredients are used in the production of cigarettes. These include humectants such as glycerol that retain moisture; sugars added to casings so that the cigarette smoke is not too alkaline or harsh; inorganic salts to alter the burning characteristics of the cigarettes; and numerous flavourants. Commonly used flavourants include liquorice, cocoa, vanillin, fruit extracts and menthol. Tobacco additives have generally been judged to be acceptable as food additives, but it is unclear whether these additives become potentially toxic in the combustion process. For example, the combustion of glycerol leads to formation of acrolein, a reactive aldehyde and respiratory irritant.[35]

Menthol has local anaesthetic qualities and is widely used as a flavourant in cigarettes, particularly in the United States. Menthol has a cooling effect on the throat. It has been speculated that menthol allows deeper inhalation and greater intake of tobacco smoke, although the evidence on this is conflicting. Mentholated cigarettes typically contain about 3 mg menthol.[36] Menthol may make cigarettes more addictive, owing to its intense sensory qualities (see Chapter 3).[37]

A number of alkylbenzene flavourants that are pulmonary irritants or are carcinogens or genotoxins have been measured in cigarettes.[38] The most common are anethole, myristicin, safrole, pulegone, piperonal and methyleugenol.

Ammonia compounds are added in significant amounts to cigarettes, up to 1.5% by weight. Ammonia in sufficient amounts can increase the pH of smoke. At higher smoke pHs, a greater proportion of nicotine in the smoke is in the un-ionized or free-base form.[39] The free-base form is more volatile than ionized nicotine, and because it is un-ionized passes through cell membranes relatively easily. The result is greater sensory impact on the throat and upper airway and potentially more rapid systemic absorption of nicotine. Recent studies of smoke pH reported values ranging from 6.0 to 7.8 for popular American cigarettes, indicating that considerable nicotine is in the free-base form in cigarette smoke.[40]

Smokeless tobacco also contains many flavourants as well as alkaline salts to facilitate buccal absorption of nicotine. Liquorice is widely used in tobacco products and can aggravate hypertension due to its mineralocorticoid actions. Users of some smokeless tobaccos consume large amounts of sodium, as much as 30–45 mmol per day, which could aggravate salt-sensitive conditions such as hypertension or heart failure.[41]

5.3.2 Carcinogens in tobacco

More than 4,500 compounds, including 69 carcinogens, have been identified in tobacco smoke, most of which are formed in the combustion process. However, there are several that are present in the tobacco before combustion, of which the tobacco-specific nitrosamines (TSNAs), N'-nitrosonornicotine (NNN) and 4-(methylnitrosamino)-1-(3-pyridyl)-1-butanone (NNK) are especially potent. TSNAs are formed from nicotine and nitrites in the curing process by microbial reduction of nitrate to nitrate. Further TSNAs are produced during combustion. NNN and NNK are present in cigarette tobacco, but the levels are quite variable, with concentrations up to 58,000 ng/g tobacco for NNN and up to 10,745 ng/g tobacco for NNK.[42] The levels of TSNAs vary by country of origin of cigarettes, with particularly high levels measured in cigarettes from India and Italy. Higher levels of TSNAs are found in dark tobacco cigarettes and cigarettes with higher nitrate content in the tobacco. New curing processes have been shown to reduce or eliminate TSNAs from cigarette tobacco.[42] Other carcinogenic volatile nitrosamines, including N-nitrosdimethylamine (NDMA), N-nitrosethylmethylamine (NEMA) and N-nitrosopyrrolidine (NPYR), are also found in cigarette tobacco.

In addition to nitrosamines, tobacco contains potential carcinogenic inorganic chemicals including arsenic, beryllium, nickel, chromium, cadmium, lead and polonium-210.

In the combustion process several potentially carcinogenic aldehydes are formed, including formaldehyde, acetaldehyde and crotonaldehyde.

Preformed NNN and NNK have been identified in cigars and bidi tobacco. Kreteks contain large amounts of eugenol, a constituent of cloves. Eugenol is a local anaesthetic but it is also cytotoxic and may contribute to pulmonary injury.

Smokeless tobacco products also contain TSNAs and other nitrosamines, as well as traces of polycyclic aromatic hydrocarbons, metals, polonium-210 and adehydes.[35] The TNSA levels in smokeless products vary widely around the world. TNSA levels have declined in recent years in Sweden and in the US, with some products containing extremely low levels;[28] however, in some countries (such as India) levels are extremely high.[43]

5.4 Addiction potential of alternative nicotine products

As described earlier, the rate of absorption of nicotine varies considerably between the many available nicotine sources. Inhalation of cigarette smoke results in the most rapid absorption of nicotine and the highest nicotine concentrations in arterial blood compared with other nicotine sources. Rapid absorption makes the product more addictive for several reasons.[44] First, high concentrations in arterial blood are delivered to the brain quickly, and these high levels produce a greater intensity of brain stimulation. Rapid delivery of high levels also allows a smoker to overcome the effects of short-term tolerance to the actions of nicotine. Since nicotine effects are perceived quickly, rapid delivery also allows the smoker to titrate the dose of nicotine to optimise effects on arousal and mood. Finally, the temporal proximity of reinforcement to drug taking is known to promote the self-administration of drugs in general. Thus the cigarette is an optimal device for promoting and sustaining nicotine addiction.

Smokeless tobacco products deliver similar daily doses of nicotine to cigarettes, but the rate of absorption is slower and the peak concentrations achieved are lower than from cigarette smoking.[25] Therefore, the intensity of positive reinforcement from a particular dose of nicotine is likely to be less from smokeless tobacco than that experienced from cigarette smoking, although physical dependence on smokeless tobacco certainly can develop. Many smokeless tobacco users become dependent and have great difficulty quitting, although some, such as professional baseball players in the US, use smokeless tobacco for putative athletic enhancement reasons.

Nicotine replacement therapies vary in the rate of nicotine absorption. Nicotine nasal spray achieves the fastest absorption, and concerns have been raised that it is likely to be the most addictive of the medicinal formulations. Nicotine gum, inhaler and lozenge have similar pharmacokinetic profiles and also provide some degree of positive reinforcement, although much less than a cigarette. Nicotine gum does have some dependence liability and some gum users have difficulty stopping gum use. Clinical trials of nicotine gum show prolonged use at 12 months after smoking cessation in 9–22% of users, and for nicotine nasal spray in 32–43% of individuals,[45] though sustained use from self-purchased products is much lower (see Chapter 4). Nicotine patches release nicotine slowly, producing little or no positive reinforcement. Dependence does not appear to be a problem with the use of nicotine patches.

There are little data on PREPs and dependence because the use of these products has been limited to small numbers of smokers.

5.5 Use of alternative nicotine sources as substitutes for cigarettes

Two questions arise in considering nicotine sources as substitutes for cigarette smoking. The first is whether these sources will be satisfying to smokers and prove to be acceptable substitutes for cigarettes. The second is whether the safety profiles are acceptable. This latter question is discussed in more detail in later chapters, though, on purely theoretical grounds, the absence of other toxins found in tobacco and particularly in its combustion products suggests strongly that these products will be far safer.

Several studies have looked at the question of whether nicotine replacements are useful for the purpose of reducing cigarette use. Some smokers are able to reduce their cigarette consumption and exposure to carbon monoxide and some other tobacco smoke toxins by using various nicotine products. Smoking reduction studies have examined the use of nicotine gum, nicotine patches, nicotine spray and nicotine oral tablets.[3,46] In general, the extent of reduction of levels of biomarkers of tobacco smoke response is less than the change expected from the magnitude of change in the self-reporting number of cigarettes smoked per day.[47,48] This may be due to misreporting of cigarette consumption, and/or to compensation by smoking each cigarette more intensively. The number of smokers who are able to substantially reduce their smoking tends to decrease over time during clinical trials, and questions remain as to whether this approach has long-term viability.

The use of smokeless tobacco as a substitute for cigarette smoking has been of considerable recent interest.[49] Smokeless tobacco does not generate combustion products, which are thought to be responsible for most of the injurious effects of cigarette smoking. The safety of smokeless tobacco products is discussed elsewhere. Smokeless tobacco has been suggested as a way to help people quit smoking, and some smokeless products have been marketed specifically for use when a person cannot smoke cigarettes. However, the long-term use of smokeless tobacco as an alternative to cigarette smoking has not yet been experimentally evaluated.

5.6 Conclusions

▶ A wide variety of nicotine products are available, delivering a range of nicotine doses. The cigarette is the most widely used product.

▶ Cigarettes deliver high doses of nicotine into the lungs, where it is absorbed rapidly and transported directly in the systemic circulation to the brain.

▶ The nicotine from cigarettes is carried in smoke which contains thousands of other chemicals, including many that are carcinogenic or otherwise toxic.

▶ Some of these toxins are present in tobacco before combustion. Most are combustion products.

▶ Smokeless tobacco also contains toxins and carcinogens, but delivers high doses of nicotine without most of the toxic components in smoke.

▶ Medicinal nicotine products deliver pure nicotine but in relatively low doses and, particularly for nicotine transdermal patches, very slowly. They do not deliver other toxic chemicals.

> The available alternative nicotine products all deliver nicotine more slowly than cigarettes, and are therefore probably less addictive.

> It is possible that alternative nicotine products could provide a safer long-term substitute for cigarette smoking. If so, this could benefit individual and public health.

References

1   Slade J, Connolly GN, Lymperis D. Eclipse: does it live up to its health claims? *Tob Control* 2002;11(Suppl 2):64–70.

2   Roethig HJ, Kinser RD, Lau RW, Walk RA, Wang N. Short-term exposure evaluation of adult smokers switching from conventional to first-generation electrically heated cigarettes during controlled smoking. *J Clin Pharmacol* 2005;45:133–45.

3   Hatsukami DK, Lemmonds C, Zhang Y *et al.* Evaluation of carcinogen exposure in people who used 'reduced exposure' tobacco products. *J Natl Cancer Inst* 2004;96:844–52.

4   Breland AB, Acosta MC, Eissenberg T. Tobacco specific nitrosamines and potential reduced exposure products for smokers: a preliminary evaluation of Advance. *Tob Control* 2003;12:317–21.

5   Stepanov I, Jensen J, Hatsukami D, Hecht SS. Tobacco-specific nitrosamines in new tobacco products. *Nicotine Tob Res* 2006;8:309–13.

6   Djordjevic MV. Nicotine dosing characteristics across tobacco products. In: Boyle P, Gray N, Henningfield J, Seffrin J, Zatonski W (eds), *Tobacco and public health: science and policy*. Oxford: Oxford University Press, 2004:181–204.

7   Wu W, Ashley DL, Watson CH. Determination of nicotine and other minor alkaloids in international cigarettes by solid-phase microextraction and gas chromatography/mass spectrometry. *Anal Chem* 2002;74:4878–84.

8   Benowitz NL. Biomarkers of cigarette smoking. In: Shopland DR, Wilkenfeld J, Henningfield J, Eriksen MP, Modell SD (eds), *The FTC cigarette test method for determining tar, nicotine and carbon monoxide yields for U.S. cigarettes*. NCI Smoking and Tobacco Control Monograph No. 7. Bethesda, MD: US National Institutes of Health, National Cancer Institute, NIH Publication No. 96-4028, August, 1996:93–111.

9   Benowitz NL. Compensatory smoking of low yield cigarettes. In: Shopland DR, Burns DM, Benowitz NL, Amacher RH (eds). *Risks associated with smoking cigarettes with low machine-measured yields of tar and nicotine*. NCI Smoking and Tobacco Control Monograph No. 13. Bethesda, MD: US National Institutes of Health, National Cancer Institute, NIH Publication No. 02-5074, October 2001:39–64.

10 Henningfield JE, Stapleton JM, Benowitz NL, Grayson RF, London ED. Higher levels of nicotine in arterial than in venous blood after cigarette smoking. *Drug Alcohol Depend* 1993;33:23–9.

11 Gourlay SG, Benowitz NL, Forbes A, McNeil JJ. Determinants of plasma concentrations of nicotine and cotinine during cigarette smoking and transdermal nicotine treatment. *Eur J Clin Pharmacol* 1997;51:407–14.

12 Hoffman D, Hoffman I. Chemistry and toxicology. In: Shopland DR, Burns DM, Hoffman D, Cummings KM, Amacher RH (eds), *Cigars – health effects and trends*. NCI Smoking and Tobacco Control Monograph No. 9. Bethesda, MD: US National Institutes of Health, National Cancer Institute, NIH Publication No. 98-4302, October 1998:55–104.

13   Fant RV, Henningfield J. Pharmacology and abuse potential of cigars. In: Shopland DR, Burns DM, Hoffman D, Cummings KM, Amacher RH (eds), *Cigars – health effects and trends*. NCI Smoking and Tobacco Control Monograph No. 9. Bethesda, MD: US National Institutes of Health, National Cancer Institute, NIH Publication No. 98-4302, October 1998:55–104.

14   Turner JA, Sillett RW, McNicol MW. Effect of cigar smoking on carboxyhaemoglobin and plasma nicotine concentrations in primary pipe and cigar smokers and ex-cigarette smokers. *Br Med J* 1977;2:1387–9.

15   Wald NJ, Idle M, Boreham J, Bailey A, Van Vunakis H. Urinary nicotine concentrations in cigarette and pipe smokers. *Thorax* 1984;39:365–8.

16   Maziak W, Ward KD, Afifi Soweid RA, Eissenberg T. Tobacco smoking using a waterpipe: a re-emerging strain in a global epidemic. *Tob Control* 2004;13:327–33.

17   Shafagoj YA, Mohammed FI, Hadidi KA. Hubble-bubble (water pipe) smoking: levels of nicotine and cotinine in plasma, saliva and urine. *Int J Clin Pharmacol Ther* 2002;40:249–55.

18   Macaron C, Macaron Z, Maalouf MT, Macaron N, Moore A. Urinary cotinine in narguila or chicha tobacco smokers. *J Med Liban* 1997;45:19–20.

19   Behera D, Uppal R, Majumdar S. Urinary levels of nicotine and cotinine in tobacco users. *Indian J Med Res* 2003;118:129–33.

20   Malson JL, Sims K, Murty R, Pickworth WB. Comparison of the nicotine content of tobacco used in bidis and conventional cigarettes. *Tob Control* 2001;10:181–3.

21   Wise MB, Guerin MR. Chemical analysis of the major constituents in clove cigarette smoke. In: Hoffman D, Harris CC (eds), *Banbury Report 23: Mechanisms in tobacco carcinogenesis*. Cold Spring Harbor, New York: Cold Spring Harbor Laboratory, 1986.

22   Smokeless Tobacco Database: Massachusetts Department of Health, 2004.

23   Determination of nicotine, pH, and moisture content of six U.S. commercial moist snuff products. Florida, January–February 1999. *MMWR Morb Mortal Wkly Rep* 1999;48:398–401.

24   Hatsukami DK, Keenan RM, Anton DJ. Topographical features of smokeless tobacco use. *Psychopharmacology (Berl)* 1988;96:428–9.

25   Benowitz NL, Porchet H, Sheiner L, Jacob P 3rd. Nicotine absorption and cardiovascular effects with smokeless tobacco use: comparison with cigarettes and nicotine gum. *Clin Pharmacol Ther* 1988;44:23–8.

26   Fant RV, Henningfield JE, Nelson RA, Pickworth WB. Pharmacokinetics and pharmacodynamics of moist snuff in humans. *Tob Control* 1999;8:387–92.

27   Wennmalm A, Benthin G, Granstrom EF *et al.* Relation between tobacco use and urinary excretion of thromboxane A2 and prostacyclin metabolites in young men. *Circulation* 1991;83:1698–704.

28   McNeill A, Bedi R, Islam S, Alkhatib MN, West R. Levels of toxins in oral tobacco products in the UK. *Tob Control* 2006;15:64–7.

29   Benowitz NL, Jacob P 3rd, Savanapridi C. Determinants of nicotine intake while chewing nicotine Polacrilex gum. *Clin Pharmacol Ther* 1987;41:467–73.

30   Benowitz NL. Clinical pharmacology of transdermal nicotine. *Eur J Pharm Biopharm* 1995;41:168–74.

31   Molander L, Lunell E, Andersson SB, Kuylenstierna F. Dose released and absolute bioavailability of nicotine from a nicotine vapor inhaler. *Clin Pharmacol Ther* 1996;59: 394–400.

32   Choi JH, Dresler CM, Norton MR, Strahs KR. Pharmacokinetics of a nicotine Polacrilex lozenge. *Nicotine Tob Res* 2003;5:635–44.

33   Fagerstrom KO, Hughes JR, Rasmussen T, Callas PW. Randomised trial investigating effect of a novel nicotine delivery device (Eclipse) and a nicotine oral inhaler on smoking behaviour, nicotine and carbon monoxide exposure, and motivation to quit. *Tob Control* 2000;9:327–33.

34   Buchhalter AR, Eissenberg T. Preliminary evaluation of a novel smoking system: effects on subjective and physiological measures and on smoking behavior. *Nicotine Tob Res* 2000;2:39–43.

35   Hoffmann D, Djordjevic MV. Chemical composition and carcinogenicity of smokeless tobacco. *Adv Dent Res* 1997;11:322–9.

36   Benowitz NL, Herrera B, Jacob P 3rd. Mentholated cigarette smoking inhibits nicotine metabolism. *J Pharmacol Exp Ther* 2004;310:1208–15.

37   Henningfield JE, Benowitz NL, Ahijevych K *et al.* Does menthol enhance the addictiveness of cigarettes? An agenda for research. *Nicotine Tob Res* 2003;5:9–11.

38   Stanfill SB, Ashley DL. Solid phase microextraction of alkenylbenzenes and other flavor-related compounds from tobacco for analysis by selected ion monitoring gas chromatography-mass spectrometry. *J Chromatogr A* 1999;858:79–89.

39   Pankow JF. A consideration of the role of gas/particle partitioning in the deposition of nicotine and other tobacco smoke compounds in the respiratory tract. *Chem Res Toxicol* 2001;14:1465–81.

40   Pankow JF, Tavakoli AD, Luo W, Isabelle LM. Percent free base nicotine in the tobacco smoke particulate matter of selected commercial and reference cigarettes. *Chem Res Toxicol* 2003;16:1014–8.

41   Benowitz NL. Sodium intake from smokeless tobacco. *N Engl J Med* 1988;319:873–4.

42   WHO International Agency for Research on Cancer. *Volume 83: Tobacco smoke and involuntary smoking.* IARC monograph on the evaluation of carcinogenic risks to humans. Lyon: IARC, 2004.

43   Stepanov I, Hecht SS, Ramakrishnan S, Gupta PC. Tobacco-specific nitrosamines in smokeless tobacco products marketed in India. *Int J Cancer* 2005;116:16–19.

44   Stitzer ML, De Wit H. Abuse liability of nicotine. In: Benowitz NL (ed), *Nicotine safety and toxicity.* New York: Oxford University Press, 1998:119–31.

45   Hughes JR. Dependence on and abuse of nicotine replacement medications: an update. In: Benowitz NL (ed), *Nicotine safety and toxicity.* New York: Oxford University Press, 1998:119–31.

46   Batra A, Klingler K, Landfeldt B *et al.* Smoking reduction treatment with 4-mg nicotine gum: a double-blind, randomized, placebo-controlled study. *Clin Pharmacol Ther* 2005;78:689–96.

47   Hatsukami DK, Henningfield JE, Kotlyar M. Harm reduction approaches to reducing tobacco-related mortality. *Annu Rev Public Health* 2004;25:377–95.

48   Hecht SS, Murphy SE, Carmella SG *et al.* Effects of reduced cigarette smoking on the uptake of a tobacco-specific lung carcinogen. *J Natl Cancer Inst* 2004;96:107–15.

49   Savitz DA, Meyer RE, Tanzer JM, Mirvish SS, Lewin F. Public health implications of smokeless tobacco use as a harm reduction strategy. *Am J Public Health* 2006;96:1934–9.

6 | The risk profile of smoked tobacco

6.1 Introduction

Cigarette smoking is established as a significant cause of death and morbidity. It is the largest identified avoidable cause of ill-health and premature death. Despite a vast accumulation of evidence since the first research studies were published in the 1950s, and public awareness of the adverse health effects, many people continue to smoke. Furthermore, smoking is emerging as a significant public health hazard in developing countries where there is potential for a large number of premature deaths.

6.2 Population trends in smoking prevalence in the United Kingdom

At the height of the smoking epidemic in England in the 1950s, as many as 80% of men and 40% of women were current smokers.[1] Since then, there has been a clear decline in the prevalence of smoking. Over the 30-year period between 1974 and 2004 the percentage of the adult population of Great Britain who smoked cigarettes almost halved (Fig 6.1). However, the rate of decline has been slower in recent years, suggesting a levelling off in prevalence.

Most people start smoking in adolescence. In Britain, more than 40% of male smokers started smoking before the age of 16 years and this has hardly changed between 1992 and 2004 (Fig 6.2).[2] The proportion of females starting to smoke before the age of 16 was smaller but has been increasing, from 28% in 1992 to 35% in 2004. This has been mirrored by a decrease in the proportion of females starting to smoke at the age of 20 or over (Fig 6.2).[2] Furthermore, the younger a person is

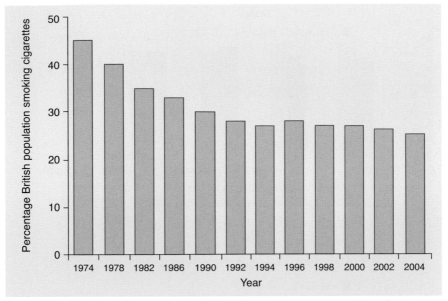

Fig 6.1 The percentage of the population of Great Britain that smoked cigarettes between 1974 and 2004. Source: General Household Survey.[2] Reproduced under the terms of the Click-Use Licence.

when they start smoking, the more cigarettes they tend to consume. Fifty-five percent of males who smoke 20 or more cigarettes each day started smoking before the age of 16 compared with 28% who smoke one to nine cigarettes each day.[2]

The prevalence of smoking in children varies significantly with age and is correlated with parental smoking. In England, the proportion of children who smoke regularly increases sharply from 1% among those aged 11 years to 20% of those aged 15 years,[3] with a greater proportion of female smokers (25% of girls compared with 16% of boys). In other countries, for example Bulgaria and the Ukraine, as many as 30% or more males aged 15 years smoke.[4] In China, an estimated 8.6% of male and 0.4% of female adolescents are current smokers, though this may reflect later uptake of smoking.[5] Children who start smoking by 15 years are likely to continue smoking well into middle age.

6.3 Trends in prevalence in other countries

While the prevalence of smoking is decreasing or stabilising in many developed countries, it is high and increasing in many countries in the developing world. Table 6.1 shows the prevalence and number of smokers in developed, developing and transitional countries (such as those in eastern Europe). In 2000, there were 1.2 billion smokers worldwide (82% male) and an estimated 1.3 billion in 2003.[6]

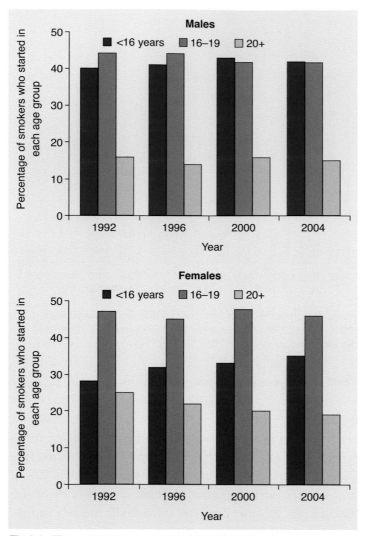

Fig 6.2 The proportion of smokers who started their habit age <16 years, 16–19 or 20 or over. Source: General Household Survey.[2] Reproduced under the terms of the Click-Use Licence.

Assuming there will be no change in prevalence, there are expected to be 1.45 billion smokers in the world in 2010, rising to 1.7 billion in 2025. As a result, the burden of morbidity and mortality, and consequent economic and healthcare costs, will increasingly be concentrated in developing countries, which have the least resources to cope with the smoking epidemic.

As described in Chapter 1, the smoking prevalence in a particular country can be modelled according to distinct stages of the smoking epidemic.[7] It may take

Table 6.1. Estimated worldwide prevalence and numbers of smokers in 2000.

Level of development	Prevalence of smoking in people aged ≥15 years (%)			Numbers of smokers (millions)*		
	Male	Female	Total	Male	Female	Total
Developed	34	21	27	115	76	191
Developing	50	7	29	810	115	924
Transitional	54	14	33	83	24	107
World	48	10	29	1,006	218	1,224

*there is some rounding. Source: Guindon *et al.*[6]

40 to 50 years for the prevalence to peak in men before smoking prevalence begins to decline. The peak and decline in women occurs several decades later than in men. Many economically developed countries are in a similar situation to the UK, but countries such as China are perhaps only beginning to reach the peak of male smoking. In 2002, there were 300 million smokers in China, with a prevalence of 66% of men and 3% of women.[8] The potential for growth in smoking among women in China and other countries predicted by the smoking epidemic model has profound implications for future public health.

6.4 Health risks associated with active smoking

Tobacco smoke contains over 3,000 chemicals, many of them toxic and/or carcinogenic, so it is not surprising that smoking has been associated with an increased risk of a wide range of diseases. Two reports on smoking in 2004 – the US Surgeon General's report and the International Agency for Research on Cancer monograph – concluded that the following disorders are caused by smoking: [9,10]

- ▸ cancer of the lung, oral cavity, pancreas, stomach, bladder, cervix, oesophagus, kidney, liver, larynx, myeloid leukaemia
- ▸ cardiovascular disease: abdominal aortic aneurysm, atherosclerosis, cerebrovascular disease, ischaemic heart disease
- ▸ respiratory disease: chronic obstructive pulmonary disease (chronic bronchitis and emphysema), pneumonia, respiratory symptoms such as cough, phlegm, wheeze and breathlessness
- ▸ reproductive effects: miscarriage and stillbirths, reduced fertility, low birth weight, complications in pregnancy such as preterm birth and premature rupture of membranes
- ▸ other effects: cataracts, hip fracture, osteoporosis, peptic ulcer.

The disorders cover a wide spectrum of illness in humans including cancer, vascular disease and respiratory disease. The link with lung cancer is well established, but it is now accepted that there are several other cancers caused by smoking.[9,10] In the UK, smoking accounts for about 30% of all cancer deaths. Table 6.2 shows the relative risks of selected fatal disorders and an estimate of the number of deaths each year in England caused by smoking.[11] Smoking also causes a range of other disorders, many that are chronic and require continuous treatment, such as diabetes and Crohn's disease. Selected non-fatal disorders, the relative risk in current smokers, and an estimate of the burden in every 100,000 adults who smoke are given in Table 6.3.[12]

The less well known effects of smoking include sight-threatening disorders such as age-related macula degeneration (AMD), cataracts and thyroid eye disease.[13–15]

Table 6.2. Estimated number of annual deaths in England caused by smoking for selected disorders.

Disorder		Relative risk in current smokers compared to never-smokers		Number of deaths due to smoking
		Men	Women	
Cancer	Lung	26.6	13.6	23,700
	Upper respiratory	10.6	6.1	600
	Oesophagus	5.3	9.3	4,000
	Bladder	2.9	1.6	1,600
	Kidney	2.8	1.3	700
	Stomach	2.1	1.2	1,500
	Pancreas	2.2	2.3	1,600
	Myeloid leukaemia	1.4	1.2	300
	Unspecified site	4.4	2.1	1,800
Respiratory	Chronic obstructive pulmonary disease	14.1	14.0	17,400
	Pneumonia	1.9 to 2.3*	2.0 to 4.6*	3,300
Circulatory	Ischaemic heart disease	1.4 to 4.2*	1.4 to 5.2*	17,700
	Stroke	1.4 to 5.1*	1.2 to 4.5*	4,500
	Aortic aneurysm	5.3	8.2	5,600
	Myocardial degeneration	2.1	1.7	200
	Atherosclerosis	1.9	2.2	200
Digestive	Stomach/duodenal ulcer	4.5	6.4	1,800
Total		–	–	86,500

* The relative risk decreases with increasing age.
Sources: adapted from two tables from the Health Development Agency (HDA) with permission.[11]
Available from National Institute for Health and Clinical Excellence website: www.nice.org.uk

Table 6.3. Summary of selected common non-fatal disorders associated with smoking.

Disorder		Gender (age if not all ages)	Approximate relative risk in current smokers compared to never-smokers	Estimate of the number of individuals developing the condition due to smoking among 100,000 smokers per year#
Gastrointestinal	Peptic ulcer	Women	2	440
		Men	2	800
	Crohn's disease	Women	4.5	190
		Men	2	80
	Gallbladder disease	Men and women	1.3	100
Skin	Psoriasis	Women	2.5	2,180
		Men	1.4	730
Neurological	Alzheimer's disease+	Men and women (≥65)	1.1	120
Eyes	Cataract extraction	Women (≥45)	1.5	70
		Men (≥45)	1.5	135
	Age-related maculopathy	Women (≥50)	2	55
		Men (≥40)	2	85
Bones	Hip fracture	Women (55–64)	1.2	10
		Women (≥85)	2.1	2,420

continued over

Table 6.3. Summary of selected common non-fatal disorders associated with smoking – *continued*.

Disorder		Gender (age if not all ages)	Approximate relative risk in current smokers compared to never-smokers	Estimate of the number of individuals developing the condition due to smoking among 100,000 smokers per year#
Fertility and pregnancy	Infertility	Women	1.4	9,500
	Miscarriage	Women	1.25	4,700
	Low birth weight (<2,500 grams)	Women	1.8	3,330
	Limb reduction defect	Women	1.5	30
Hormonal	Type II diabetes mellitus	Women (30–55)	1.2	35
		Men (≥40)	2	240
Teeth	Periodontitis	Men and women	4	17,140

Risk difference between smokers and non-smokers. Based on the incidence of the disorder (per 100,000 per year), except psoriasis and periodontitis (based on the prevalence) and miscarriage, limb reduction defects and birth weight (based on 100,000 births).

+ Included for interest, since although there is a suggestion of an association there may not be one at all.

Reproduced from Hackshaw.[12]

Age-related macular degeneration is the most common cause of blindness in most developed countries. In the UK, an estimated 17,800 people are blind because of smoking.[16]

In the UK, about 115,000 people died from smoking in 2000, compared with 512,000 in the USA (Table 6.4).[17] This represents about a fifth of all deaths in each of those two countries. There could be more than one billion deaths from smoking in the 21st century if current trends continue.[18]

Table 6.4. Number and percentage of deaths caused by smoking in selected developed countries in 2000.

Country	Total number of deaths ('000)	Number attributable to smoking ('000)	Percentage of all deaths attributable to smoking
Canada	218	44	20
France	531	60	11
Germany	839	109	13
Italy	560	80	14
Japan	962	113	12
Spain	360	451	13
UK	608	115	19
USA	2,403	512	21

Reproduced from Peto et al.[17]

Worldwide, smoking is responsible for a massive 4.8 million deaths each year (range 3.9 to 5.9 million), and more than half of these occur in developing countries (Table 6.5).[19] This is equivalent to about one death every seven seconds. As smoking declines in developed countries over the forthcoming years, the burden of smoking-related illness will shift to the poorer countries. The full burden of smoking-related disease and death in large countries such as China and India will not arise until several decades from now.

The full effects of smoking on mortality can be seen in long-term studies of smokers, including one study of British doctors.[20] This study established that about half of continuing smokers die prematurely from their habit, though it may be that as many as two thirds of all smokers die early.[20] A quarter of these deaths occur in middle age (35–69 years) and life expectancy is reduced, on average, by about 10 years (Fig 6.3).[20] It is only recently that the scale of the risk to smokers has been fully realised and quantified. Many smokers may be unaware of the magnitude of this effect.

Smoking is one of the main causes of social inequalities in health, especially in developed countries. Smoking accounts for about half the number of deaths in the

Table 6.5. Estimated worldwide mortality (in thousands) attributable to smoking by cause of death in 2000.

Cause	Men 30–69 yrs	Men 70+ yrs	Women 30–69 yrs	Women 70+ yrs	Total
Lung cancer	398	294	77	79	848
Cancer of mouth, oropharynx or oesophagus	152	66	17	15	250
Other cancer	195	135	17	24	371
Chronic obstructive pulmonary disease	269	433	86	178	966
Other respiratory disease	274	93	34	32	433
Cardiovascular disease	848	476	143	223	1,690
Other causes	145	57	36	35	273
Total deaths	2,280	1,556	410	587	4,831

Reproduced from Ezzati *et al* with permission from Elsevier, copyright 2003.[19]

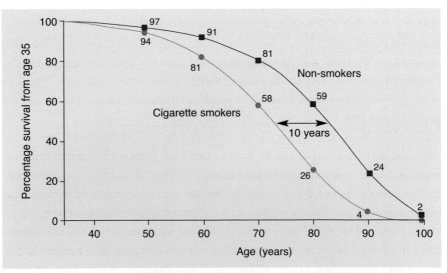

Fig 6.3 Survival from age 35 for continuing cigarette smokers and lifelong non-smokers among UK male physicians born 1900–1930. Reproduced from Doll *et al* with permission from the BMJ Publishing Group.[20]

lowest socio-economic group (Fig 6.4).[21] In England and Wales, there is a five-fold increase in smoking-related deaths among people in the lowest socio-economic group compared with those in the highest. Most of those who died aged 35–69 would have survived to 70 years and above if they had not smoked. The effect is similar in other countries. The effects of smoking on social inequalities in health are discussed in more detail in Chapter 10.

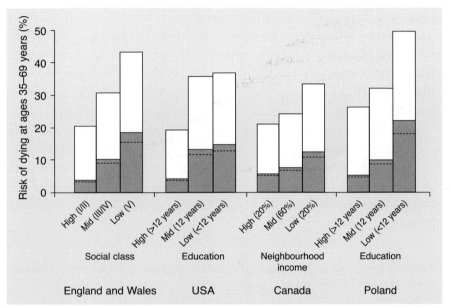

Fig 6.4 Bar chart showing social inequalities in male mortality. Each bar shows the risk of a man aged 35 years old in 1996 dying from any cause between the age of 35 and 69 years. The shaded section indicates how much of this risk is due to smoking. The shaded part of the bar up to the dashed lines represents the proportion who would have lived beyond 70 years had they not smoked.

6.5 Health risks associated with passive smoking

The evidence for harm caused by passive smoking (exposure to environmental tobacco smoke, involuntary smoking or second-hand smoking) emerged in the 1970s. The smoke breathed in by a non-smoker contains the same range of toxic substances as that inhaled by active smokers, and typically delivers about 1% of the quantity of these substances inhaled by active smokers.[22] It is, therefore, plausible that passive smoking causes the same disorders that affect smokers, but at a lower risk. There is direct evidence from observational studies of adult non-smokers that passive smoking increases the risk of developing lung cancer, ischaemic heart disease, chronic obstructive pulmonary disease (COPD) and stroke, with percentage increases in risk of 24%, 30%, 25% and 45% respectively.[22,23] These represent important increases in risk, particularly in countries where the smoking prevalence is high such as China, and one in eight non-smoking men and half of non-smoking women are exposed to passive smoke at home, and about a quarter are exposed at work.[24] There is, therefore, the potential for a large number of non-smokers, as well as smokers, in China to develop lung cancer, heart disease and COPD in the future due to passive smoking exposure, as a consequence of the high prevalence of smoking.

Health concerns about passive smoking tend to focus on lung cancer because this disease is largely specific to smoking. However, ischaemic heart disease (IHD) is much more common in non-smokers, so although the effect of passive smoking on IHD risk is much weaker than on lung cancer, the number of individuals affected by IHD is much greater. This fact is often overlooked.

In the European Union, there are over six times more IHD deaths than lung cancer deaths because of passive smoking (Table 6.6).[23] Strokes caused by smoking are estimated to account for several thousand premature deaths each year. In total, an estimated 19,242 non-smokers in the EU die each year from these four causes by breathing other people's tobacco smoke – equivalent to one death every 27 minutes

Table 6.6. Estimated number of deaths in 2002 among non-smokers due to passive smoking in 24 EU countries.#

Cause of death	Exposure at home, adults			Exposure at work		Total (home plus workplaces)*
	24<65	65+	All	All work-places	Hospitality industry	
Lung cancer	403	629	1,032	521	16	1,553
Ischaemic heart disease	1,781	6,977	8,758	1,481	48	10,239
Stroke	729	4,954	5,683	596	19	6,279
Chronic obstructive pulmonary disease	155	815	970	201	6	1,171
Total*	3,068	13,375	16,443	2,799	89	19,242

* there is some rounding
# Austria, Belgium, Czech Republic, Denmark, Estonia, Finland, France, Germany, Greece, Hungary, Ireland, Italy, Latvia, Lithuania, Luxembourg, Malta, Netherlands, Poland, Portugal, Slovakia, Slovenia, Spain, Sweden and UK.
Reproduced from Jamrozik with permission from the European Respiratory Society Journals Ltd.[23]

6.6 Effects of smoking cessation

In the same way that the effects of smoking were not fully realised until a large number of smokers had been followed for many years, the effect of giving up has also only recently been fully quantified. In the British Doctors Study, a large proportion of participants were long-term smokers but then quit in the 1950s and 1960s.[20] From this, it is possible to quantify accurately the benefits of stopping in a group of people in whom the risk of dying because of smoking was substantial before they had quit. People who give up smoking can benefit from a significant reduction in risk for all fatal disorders, though the risk does not decrease to the same level in never-smokers (Table 6.7).[20]

There is a clear relationship between the age when a smoker quits and the risk of dying prematurely (Fig 6.5).[20] If a smoker quits when aged about 30 years, they avoid almost all the risk of premature death from smoking. Someone aged 50 years could halve the chance of premature death, and even quitting at 60 years could reduce a smoker's risk and gain them, on average, three years of life expectancy.

While stopping smoking undoubtedly has benefits on overall mortality, the degree of benefit varies for different disorders. It may take 20–30 years before the risk of lung cancer in an ex-smoker is about 80% lower than the risk in continuing smokers.[1] After about 15 years, the risk could be halved. In contrast, an

Table 6.7. Average percentage change in risk among all ex-smokers compared to that in current smokers or never-smokers.*

Cause of death	Percentage increase in risk in ex-smokers compared to never-smokers (%)	Percentage decrease in risk in ex-smokers compared to current smokers (%)
Lung cancer	300	73
Cancer of the mouth, pharynx, larynx or oesophagus	189	57
All other cancers	11	21
Chronic obstructive pulmonary disease	482	59
Other respiratory disease	34	29
Ischaemic heart disease	23	24
Cerebrovascular disease	16	26
Other vascular disease	24	32
Other medical conditions	9	29
All causes	25	32

* based on the age-standardised death rate per 1,000 men/year.
Reproduced from Doll *et al* with permission from the BMJ Publishing Group.[20]

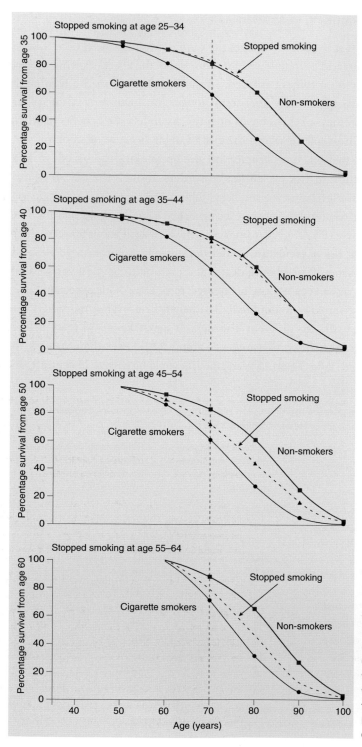

Fig 6.5 The effect on survival of stopping smoking cigarettes at age 25–34 (effect from age 35), age 35–44 (effect from age 40), age 45–54 (effect from age 50), and age 55–64 (effect from age 60). Reproduced from Doll *et al* with permission from the BMJ Publishing Group.[20]

estimated third of the risk of developing coronary heart disease could be avoided after only two years of cessation and almost all of the risk is eliminated after 10 years.[25] Similarly, there is evidence that most of the risk of having a stroke disappears between two and four years after quitting.[26] Smoking cessation is also of benefit to people who have already developed disease. Smokers with coronary heart disease who quit can reduce their risk of dying prematurely by 36%, compared with continuing smokers.[27]

6.7 Conclusions

▶ Smoking currently kills 5 million people each year.

▶ In the 20th century there were an estimated 100 million premature deaths attributable to smoking. If current smoking patterns continue there could be more than 1 billion deaths in the 21st century.

▶ A large proportion of the population in many countries still take up smoking when young and continue their habit into middle and old age. Of all those who die from smoking worldwide, half are in developing countries, but this proportion is likely to increase unless the smoking epidemic can be halted.

▶ Passive smoking is also a major avoidable cause of death and disability.

▶ Stopping smoking is highly effective, even in older smokers, and generates immediate benefits to health.

▶ To avoid a greater public health disaster in the current century more efforts should be made to prevent non-smokers from starting to smoke, and to encourage smokers to quit.

References

1   Peto R, Darby S, Deo H *et al.* Smoking, smoking cessation, and lung cancer in the UK since 1950: combination of national statistics with two case-control studies. *BMJ* 2000; 321:323–9.

2   Office for National Statistics. *General Household Survey, 2004.* www.statistics.gov.uk/StatBase/Product.asp?vlnk=5756 (accessed 7 August 2007).

3   National Centre for Social Research. *Drug use, smoking and drinking among young people in England in 2005.* NHS Health and Social Care Information Centre, Public Health Statistics, 2006.

4   Mackay J, Eriksen M, Shafey O. *The tobacco atlas,* 2nd edn. American Cancer Society, 2006.

5   Honghuan Y. The epidemic of tobacco use in China. In: Boyle P, Gray N, Henningfield J, Seffrin J, Zatonski W (eds), *Tobacco: the public health disaster of the twentieth century.* Oxford: Oxford University Press, 2004.

6   Guindon GE, Boisclair D. *Past, current, and future trends in tobacco use.* The World Bank, February 2003. www.worldbank.org/tobacco/pdf/Guindon-Past,%20current-%20whole. pdf (accessed 7 August 2007).

7   Lopez AD, Collishaw NE, Piha T. A descriptive model of the cigarette epidemic in developed countries. *Tob Control* 1994;3:242–7.

8   Sung H-Y, Wang L, Jin S, Hu T-W, Jiang Y. Economic burden of smoking in China, 2000. *Tob Control* 2006;15(Suppl 1):5–11.

9   US Department of Health and Human Services. *The health consequences of smoking.* Report of the US Surgeon General. Atlanta, GA: US Department of Health and Human Services, Centers for Disease Control and Prevention, National Center for Chronic Disease Prevention and Health Promotion, Office on Smoking and Health, 2004.

10   Vineis P, Alavanja M, Buffler P *et al*. Tobacco and cancer: recent epidemiological evidence. *J Natl Cancer Inst* 2004;96:99–106.

11   Health Development Agency. *The smoking epidemic in England.* London: Health Development Agency, 2004. Available from www.nice.org.uk

12   Hackshaw AK. Tobacco and other diseases. In: Boyle P, Gray N, Henningfield J, Seffrin J, Zatonski W (eds), *Tobacco: the public health disaster of the twentieth century.* Oxford University Press, 2004.

13   Thornton J, Kelly SP, Edwards R *et al*. Smoking and age-related macular degeneration: a review of association. *Eye* 2005;19:935–44.

14   Kelly SP, Thornton J, Edwards R, Sahu A, Harrison R. Smoking and cataract: review of causal association. *J Cataract Refract Surg* 2005;31:2395–2404.

15   Thornton J, Harrison R, Kelly SP, Edwards R. Cigarette smoking and thyroid eye disease: a systematic review. *Eye,* 15 September 2006 [Epub ahead of print].

16   Kelly SP, Thornton J, Lyratzopoulos G, Edwards R, Mitchell P. Smoking and blindness. Strong evidence for the link, but public awareness lags. *BMJ* 2004;328:357–8.

17   Peto R, Lopez AD, Boreham J, Thun M. Mortality from smoking in developed countries 1950–2000, June 2006. www.ctsu.ox.ac.uk/~tobacco/ (accessed 7 August 2007).

18   Peto R, Lopez AD. Future worldwide health effects of current smoking patterns. In: Koop CE, Pearson CE, Schwarz MR (eds), *Critical issues in global health.* San Francisco: Jossey-Bass, 2001.

19   Ezzati M, Lopez AD. Estimates of global mortality attributable to smoking in 2000. *Lancet* 2003;362:847-52.

20   Doll R, Peto R, Boreham J, Sutherland I. Mortality in relation to smoking: 50 years' observations on male British doctors. *BMJ* 2004;328:1519–27.

21   Jha P, Peto R, Zatonski W, Boreham J, Jarvis MJ, Lopez AD. Social inequalities in male mortality, and in male mortality from smoking: indirect estimation from national death rates in England and Wales, Poland, and North America. *Lancet* 2006;368:367–70.

22   Royal College of Physicians. *Going smoke-free: the medical case for clean air in the home, at work or in public places.* Report of the Tobacco Advisory Group of the Royal College of Physicians. London: RCP, 2005.

23   Jamrozik K. An estimate of deaths attributable to passive smoking in Europe. In: *Smoke free partnership. Lifting the smokescreen: 10 reasons for a smoke free Europe.* Brussels: ERSJ Ltd, 2006:17–41.

24   Gu D, Wu X, Reynolds K *et al*. Cigarette smoking and exposure to environmental tobacco smoke in China: the International Collaborative Study of Cardiovascular Disease in Asia. *Am J Pub Health* 2004;94:1972–6.

25   Kawachi I, Colditz G, Stampfer MJ *et al*. Smoking cessation and time course of decreased risks of coronary heart disease in middle-aged women. *Arch Intern Med* 1994;154:169–75.

26   Kawachi I, Colditz G, Stampfer MJ *et al*. Smoking cessation and decreased risk of stroke in women. *JAMA* 1993;269:232–6.

27   Critchley J, Capewell S. Smoking cessation for the secondary prevention of coronary heart disease. *Cochrane Database of Syst Rev* 2003;(4):CD003041.

7 | The risks of medicinal nicotine

7.1 Introduction

Medicinal nicotine, commonly referred to as nicotine replacement therapy or NRT, has been available as a therapy for smokers trying to quit smoking for over 20 years. Nicotine replacement therapy has been used extensively in the UK, particularly since the relaxation of general sales restrictions in 1999 and 2001 and since all NRT products became available on reimbursable National Health Service prescriptions in 2001 (Fig 7.1).[1] Clinical experience with NRT is extensive and provides strong evidence that NRT is a safe and well tolerated treatment.[2] Nicotine replacement therapy may have adverse effects, however, including local effects occurring at the point of nicotine absorption, which are relatively specific to the mode of delivery, and systemic effects that probably apply to all products. In general, the local effects tend to be mild and transient, and information about these is readily available from the large number of clinical trials of NRT products. In contrast, information on potentially important systemic effects, such as precipitation of acute cardiovascular events, is generally less easily available because these events are rare and their risk is difficult to assess in clinical trials. However, observational data have been helpful in assessing the likely magnitude of these risks.

7.2 Local effects

The frequency and severity of local adverse effects varies with the route of administration and the dose of NRT used.

7.2.1 Nicotine transdermal patches

Local skin reactions to nicotine transdermal patches, including itching and skin irritation, are common. For example, in one trial of 3,575 smokers, skin itching

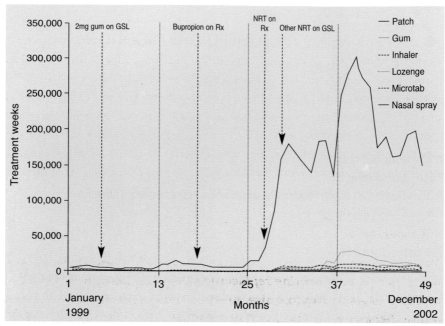

Fig 7.1 Monthly prescriptions of nicotine replacement therapy used for smoking cessation in the UK 1999 to 2002 (month 1 being January 1999). Arrows indicate introduction of policy. GSL = general sale list (available over-the-counter); Rx = on prescription. Reproduced from West *et al* with permission from the BMJ Publishing Group.[1]

occurred in 14.4% of users of a 25 mg patch; in 12.9% with a 15 mg patch; and in 5% of users with a placebo.[3] In a trial using a 21 mg patch, adverse skin reactions occurred in 18.5% in the active treatment group and in 6.9% in the placebo group.[4] In most cases skin reactions are short-lived and the risk can be reduced by varying the site used for patch application. Further evidence of the transient nature of this response is provided by an observational survey of patch users two weeks after treatment, in which only 9.5% of users recalled a skin reaction.[5]

7.2.2 Gum, lingual tablets and lozenges

Nicotine gum and other oral nicotine products can cause dyspepsia and gastro-intestinal discomfort. For example, in a clinical trial of 4 mg gum involving 364 people, 16% of people using gum reported dyspepsia compared with 2% in the placebo group.[6] Similar figures have been published in other trials.[2] In one head-to-head comparison between 4 mg gum and 4 mg lozenge, the risk of dyspepsia or nausea was similar for the two preparations.[7] Gum has also been reported to cause gingival pain, toothache and hiccups.[2,8]

7.2.3 Inhalators and nasal spray

The main local adverse effects of nicotine inhalators and nasal sprays are throat or nose irritation and cough.[2] In one placebo-controlled trial of 400 participants, throat irritation occurred in 7% of nicotine inhalator users, compared with 2% with a placebo.[9] The equivalent figures for cough were similar at 6.5% and 2% respectively. Other trials involving inhalators have reported lower adverse event rates.[10] Nasal sprays are generally well tolerated. For example, in a trial of 227 smokers randomised to receive four weeks of treatment with a nasal spray or a matched placebo, only two subjects (one in the active arm and one in the placebo arm) were advised to stop using their spray because of local side effects.[11]

7.3 Systemic adverse effects of nicotine replacement therapy

Most users and healthcare workers are not greatly worried by transient local effects of NRT, but would be concerned if NRT was found to have more serious systemic adverse effects. Particular concerns include a possible increase in incidence of cardiovascular events and effects during pregnancy on fetal development and the risk of congenital abnormalities.

7.3.1 Cardiovascular risk and sudden death

Nicotine causes catecholamine release and vasoconstriction and, therefore, an increase in blood pressure and heart rate. It is therefore plausible that NRT could precipitate acute cardiovascular events, particularly in susceptible people. However, the peak venous levels of nicotine occurring with NRT are typically 30–50% lower than those from cigarettes, and around five times lower than the peak arterial levels that cigarettes deliver, suggesting that NRT should be safer than continuing to smoke (Fig 7.2).[12] Furthermore, in one study of 12 healthy smokers, doses of NRT of up to 63 mg per day (that is, around two to three times higher than the typical daily dose) combined with smoking did not affect heart rate or blood pressure.[13]

Concerns about cardiovascular side effects grew following case reports of acute myocardial infarction in people using NRT.[14] However, such studies provide little evidence about causation, and the observed events may have been coincidental. Data from observational epidemiological studies and controlled trials are required to test the hypothesis that NRT causes cardiovascular side effects.

In general, clinical trials using healthy people have insufficient statistical power to detect an adverse impact of NRT on the risk of having an acute cardiovascular event. For example, in the CEASE study (one of the largest clinical

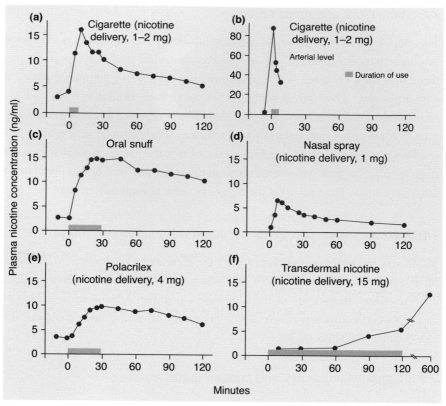

Fig 7.2 Plasma nicotine concentrations from different sources. Reproduced from
Henningfield JE. Nicotine medications for smoking cessation. *N Engl J Med*
1995;333:1196–203.[12] Copyright1995 © Massachusetts Medical Society. All rights reserved.

trials) 3,575 smokers were randomised to receive a placebo, a 15 mg patch or a
25 mg patch for eight or 22 weeks.[3] Four myocardial infarctions occurred, but
only one of these took place during active NRT treatment. Two of the other cases
occurred after participants had completed their course of NRT, and one in a
participant in the placebo group. Most other clinical trials involving NRT involve
considerably smaller numbers of participants. However, a meta-analysis of 35
clinical trials involving more than 9,000 participants found no evidence of an
increase in the incidence of acute cardiovascular events with the use of NRT.[15]

The cardiovascular effects of NRT have been investigated in studies of individuals with established ischaemic heart disease. Such individuals could be at
increased risk of cardiovascular side effects due to the presence of pre-existing
heart disease. In one study of 36 patients with known ischaemic heart disease,
the use of nicotine patches (14 mg and 21 mg) reduced the extent of exercise-
induced myocardial ischaemia assessed by thallium scanning.[16] The authors

suggested that the mechanism for this unexpected beneficial effect may be reduced cigarette consumption. A larger study assessed the effect of NRT on a composite endpoint comprising death, myocardial infarction, cardiac arrest, worsening angina requiring hospital admission, arrhythmia, or congestive heart failure in 584 smokers with a past history of cardiovascular disease randomised to receive patches (21 mg for six weeks, then 14 mg for two weeks, then 7 mg for two weeks) or a placebo. By week 14, this composite outcome occurred in 5.4% of those who had received NRT, and 7.9% of those who received a placebo. In addition, there was no significant increase in other cardiovascular adverse effects in the actively treated group.[17] Similar reassurance comes from trials in people with coronary artery disease which have included 24-hour ECG monitoring and/or exercise testing and which demonstrate that NRT does increase the risk of having sub-clinical ST segment depression/ischaemic episodes.[18,19]

Data on the risks of NRT use in the general population are available from an analysis of observational data from 33,247 UK primary care patients on the Health Improvement Network (a primary care consultation record database) who were prescribed NRT between June 1985 and November 2003. The study compared the risk of events in subjects in the 56 days before and after the first NRT prescription, using the remaining available person-time as the baseline. There was no evidence of an increased risk of acute myocardial infarction, acute stroke or death in the 56 days after the first prescription for NRT (Figs 7.3 and 7.4).[20] The estimated risk of cardiac or cerebrovascular events was actually higher than average in the few weeks before prescription, suggesting that general practitioners may have tended to prescribe NRT to people shortly after acute cardiovascular events, when the risk of further events would be expected to be particularly high (Figs 7.3 and 7.4). Alternatively or additionally, the reduced risk may simply be attributable to smoking cessation. The death rate during the first 56 days after starting treatment with NRT was, if anything, slightly lower than that during subsequent follow-up (hazard ratio 0.86; 95% confidence interval 0.60 to 1.23).

Overall, therefore, the clinical trial and observational data indicate that, in relation to cardiovascular outcomes, NRT is safe and specifically does not increase the incidence of acute cardiovascular events or of sudden death in healthy volunteers, the general population or patients with pre-existing cardiovascular disease.

7.3.2 Effects on the fetus

There is abundant evidence that smoking during pregnancy is associated with higher levels of miscarriage, intrauterine growth retardation, preterm birth and perinatal mortality.[21] Despite this, there have been few studies of the use of NRT

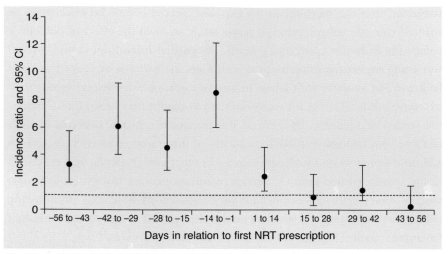

Fig 7.3 Temporal relation between relative incidence of acute myocardial infarction and first prescription for NRT in UK primary care data. Dotted line indicates an incidence ratio of 1 (no effect). Reproduced from Hubbard *et al* with permission from the BMJ Publishing Group Ltd.[20]

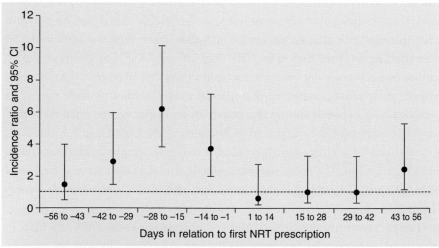

Fig 7.4 Temporal relation between relative incidence of acute stroke and first prescription for NRT in UK primary care data. Dotted line indicates an incidence ratio of 1 (no effect). Reproduced from Hubbard *et al* with permission from the BMJ Publishing Group Ltd.[20]

during pregnancy, mainly because of the logistical difficulties of undertaking drug intervention trials in pregnant women. In the only published trial of NRT in pregnant women, 250 healthy pregnant smokers were randomised to receive either a nicotine patch (15 mg patch for eight weeks followed by 10 mg patch for three weeks) or a matched placebo.[22] Adherence to the trial medications was low in this study, but no serious adverse events were recorded, and the mean birth weight of children born to women who were given NRT, in an intention-to-treat analysis, was 186 g higher than that in the placebo group. An observational study has looked at the safety of NRT using data from the Danish Nation Birth Cohort,[23] which contains information for 76,768 first born singletons born between January 1997 and December 2003. The proportion of women smoking during pregnancy was 27%, and overall the risk of congenital malformations was no different for children born to either smokers or non-smokers. A total of 231 women used NRT during the first trimester of pregnancy, and among the children born to these women there were 19 congenital malformations. This represented an increase in risk in comparison with non-smokers on the borderline of statistical significance (prevalence ratio 1.61; 95% confidence interval 1.01 to 2.58). Among these 19 congenital malformations, 14 were musculoskeletal (including seven cases of hip dislocation) and, on the basis of EUROCAT definitions (www.eurocat.ulster. ac.uk), six were classified as major abnormalities. The risk of major abnormalities was not increased in relation to the unexposed people (prevalence ratio 1.13, 95% confidence interval 0.62 to 2.07).

Overall, the available data on the safety of NRT during pregnancy are limited, and more clinical trials and post-marketing surveillance studies are needed.[24] However, the data available suggest that nicotine does not reduce birth weight, and is not a cause of serious developmental abnormalities. The unexpected finding of an increase in the incidence of non-major musculoskeletal abnormalities is based on a small number of events in a single study. However, it requires further investigation.

7.3.3 Breastfeeding

Nicotine enters breast milk,[25] and in a study of 15 breastfeeding mothers who smoked an average of 17 cigarettes per day, the nicotine level in breast milk during periods of smoking was similar to that during the use of a 21 mg patch.[26] When the mothers were given 14 mg or 7 mg patches, the nicotine levels in breast milk were below the levels during normal smoking. The impact of breast milk nicotine exposure on the infant is not known, but is clearly likely to be no greater than that arising from active smoking by the mother.

7.3.4 Other potential health effects

At present there are no data from either clinical trials or from observational studies to suggest that the use of NRT causes or exacerbates any of the other major health problems caused by cigarette smoking, such as chronic obstructive pulmonary disease, lung cancer, pneumonia or peripheral vascular disease. There are theoretical concerns arising from experiments in mice that suggest that when nicotine comes into contact with tumours implanted in mucosal surfaces, it may promote tumour growth.[27] Recent work to assess effects on angiogenesis in animals also raises the possibility that nicotine could promote tumour growth and atherosclerosis. Such effects may be more relevant in association with sustained long-term use of NRT.[28,29] However, there is no evidence that this theoretical risk derived from animal studies translates into an increase in cancer risk or tumour growth in humans.

7.4 Conclusions

▸ Extensive experience with nicotine replacement therapy in clinical trial and observational study settings demonstrates that medicinal nicotine is a very safe drug.

▸ Adverse effects are primarily local and specific to the mode of delivery used.

▸ NRT does not appear to provoke acute cardiovascular events, even in people with pre-existing cardiovascular disease.

▸ There is no direct evidence that NRT therapy is carcinogenic or influences the risk of other common smoking-related diseases in humans.

▸ Evidence on the safety of NRT during pregnancy is limited, but suggests that NRT does not increase the risk of major developmental anomalies or reduce birth weight. However, NRT may increase the risk of minor musculoskeletal anomalies. Further evidence on these effects is needed.

▸ Evidence on the safety of long-term use of NRT is lacking, but there are no grounds to suspect appreciable long-term adverse effects on health.

▸ In any circumstance, the use of NRT is many orders of magnitude safer than smoking.

References

1   West R, DiMarino ME, Gitchell J, McNeill A. Impact of UK policy initiatives on use of medicines to aid smoking cessation. *Tob Control* 2005;14:166–71.

2   Silagy C, Lancaster T, Stead L, Mant D, Fowler G. Nicotine replacement therapy for smoking cessation. *Cochrane Database Syst Rev* 2004;(3):CD000146.

3   Tonnesen P, Paoletti P, Gustavsson G *et al.* Higher dosage nicotine patches increase one-year smoking cessation rates: results from the European CEASE trial. Collaborative European Anti-Smoking Evaluation. European Respiratory Society. *Eur Respir J* 1999; 13:238–46.

4   Jorenby DE, Leischow SJ, Nides MA *et al*. A controlled trial of sustained-release bupropion, a nicotine patch, or both for smoking cessation. *N Engl J Med* 1999;340:685–91.

5   Hasford J, Fagerstrom KO, Haustein KO. A naturalistic cohort study on effectiveness, safety and usage pattern of an over-the-counter nicotine patch. Cohort study on smoking cessation. *Eur J Clin Pharmacol* 2003;59:443–7.

6   Batra A, Klingler K, Landfeldt B *et al*. Smoking reduction treatment with 4-mg nicotine gum: a double-blind, randomized, placebo-controlled study. *Clin Pharmacol Ther* 2005; 78:689–96.

7   Marsh HS, Dresler CM, Choi JH *et al*. Safety profile of a nicotine lozenge compared with that of nicotine gum in adult smokers with underlying medical conditions: a 12-week, randomized, open-label study. *Clin Ther* 2005;27:1571–87.

8   Moolchan ET, Robinson ML, Ernst M *et al*. Safety and efficacy of the nicotine patch and gum for the treatment of adolescent tobacco addiction. *Pediatrics* 2005;115:e407–14.

9   Bolliger CT, Zellweger JP, Danielsson T, van Biljon X, Robidou A, Westin A *et al*. Smoking reduction with oral nicotine inhalers: double blind, randomised clinical trial of efficacy and safety. *BMJ* 2000;321:329–33.

10 Bohadana A, Nilsson F, Rasmussen T, Martinet Y. Nicotine inhaler and nicotine patch as a combination therapy for smoking cessation: a randomized, double-blind, placebo-controlled trial. *Arch Intern Med* 2000;160:3128–34.

11 Sutherland G, Stapleton JA, Russell MA *et al*. Randomised controlled trial of nasal nicotine spray in smoking cessation. *Lancet* 1992;340:324–9.

12 Henningfield JE. Nicotine medications for smoking cessation. *N Engl J Med* 1995;333: 1196–203.

13 Zevin S, Jacob P, III, Benowitz NL. Dose-related cardiovascular and endocrine effects of transdermal nicotine. *Clin Pharmacol Ther* 1998;64:87–95.

14 Dacosta A, Guy JM, Tardy B *et al*. Myocardial infarction and nicotine patch; a contributing or causative factor? *Eur Heart J* 1993;14:1709–11.

15 Greenland S, Satterfield MH, Lanes SF. A meta-analysis to assess the incidence of adverse effects associated with the transdermal nicotine patch. *Drug Safety* 1998;18:297–308.

16 Mahmarian JJ, Moye LA, Nasser GA *et al*. Nicotine patch therapy in smoking cessation reduces the extent of exercise-induced myocardial ischaemia. *J Am Coll Cardiol* 1997;30: 125–30.

17 Joseph AM, Norman SM, Ferry LH *et al*. The safety of transdermal nicotine as an aid to smoking cessation in patients with cardiac disease. *New Engl J Med* 1996;335:1792–8.

18 Working group for the study of Transdermal Nicotine in Patients with Coronary Artery Disease. Nicotine replacement therapy for patients with coronary heart disease. *Arch Int Med* 1994;154:989–95.

19 Tzivoni D, Keren A, Meyler S *et al*. Cardiovascular safety of transdermal nicotine patches in patients with coronary artery disease who try to quit smoking. *Cardiovasc Drugs Ther* 1998;12:239–44.

20 Hubbard R, Lewis S, Smith C *et al*. Use of nicotine replacement therapy and the risk of acute myocardial infarction, stroke, and death. *Tob Control* 2005;14:416–21.

21 Coleman T, Britton J, Thornton J. Nicotine replacement therapy in pregnancy. *BMJ* 2004;328:965–6.

22 Wisborg K, Henriksen TB, Jespersen LB, Secher NJ. Nicotine patches for pregnant smokers: a randomized controlled study. *Obstet Gynecol* 2000;96:967–71.

23 Morales-Suarez-Varela MM, Bille C, Christensen K, Olsen J. Smoking habits, nicotine use, and congenital malformations. *Obstet Gynecol* 2006;107:51–7.

24   Dempsey D, Jacob P, III, Benowitz NL. Accelerated metabolism of nicotine and cotinine in
      pregnant smokers. *J Pharmacol Exp Ther* 2002;301:594–8.

25   Dahlstrom A, Ebersjo C, Lundell B. Nicotine exposure in breastfed infants. *Acta Paediatr*
      2004;93:810–6.

26   Ilett KF, Hale TW, Page-Sharp M *et al.* Use of nicotine patches in breast-feeding mothers:
      transfer of nicotine and cotinine into human milk. *Clin Pharmacol Ther* 2003;74:516–24.

27   Shin VY, Wu WK, Ye YN *et al.* Nicotine promotes gastric tumor growth and neo-
      vascularization by activating extracellular signal-regulated kinase and cyclooxygenase-2.
      Carcinogenesis 2004;25:2487–95.

28   Cooke JP, Bitterman H. Nicotine and angiogenesis: a new paradigm for tobacco-related
      diseases. *Ann Med* 2004;36:33–40.

29   Heeschen C, Jang JJ, Weis M *et al.* Nicotine stimulates angiogenesis and promotes tumor
      growth and atherosclerosis. *Nat Med* 2001;7:833–9.

8 | The risk profile of smokeless tobaccos

8.1 Introduction

In the context of the history of tobacco use, as outlined in Chapter 1, cigarette smoking is a recent phenomenon. Tobacco has been used in human societies for at least 2,000 years, but the epidemic of cigarette smoking, which at the end of the 20th century involved over one billion people (47% of all adult men and 12% of all adult women in the world),[1] originated little more than a century ago.

Before the emergence of cigarette smoking in 'mature' tobacco markets such as northern Europe, North America and Australia, smokeless tobacco products (which in Europe included nasal snuff) were initially the most widely used, while pipe and cigar smoking became popular in the 19th century. However, in the latter half of the 20th century, tobacco consumption in these populations reduced steadily, largely as a result of increasing awareness of the harmful effects of smoking on health.[2] The pattern of consumption of these different products in the United States during the past 120 years illustrates these trends (Fig 8.1) and is typical of experience in many mature tobacco market countries.[3]

Today, manufactured cigarettes are the dominant nicotine product consumed in the European Union and most other parts of the world. Other forms of smoked tobacco (cigars, pipes, roll-your-own, bidis and kreteks) are also used in many countries. The combustion of tobacco and inhalation of the resulting smoke causes the rapid absorption of nicotine along with thousands of other toxins in the lungs. This high delivery of toxins to the lungs of tobacco smokers has caused a global epidemic of lung disease, primarily lung cancer and chronic obstructive pulmonary disease (COPD).

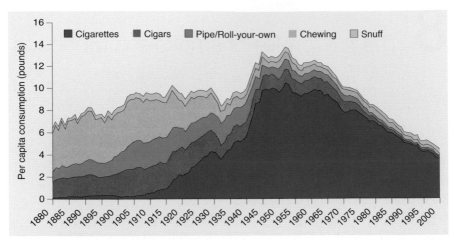

Fig 8.1 Per capita consumption of various tobacco products (in pounds) in the United States, 1880–2000. Census note: among persons >18 years old. Beginning in 1982, fine-cut chewing tobacco was reclassified as snuff. Reproduced from *Tobacco situation and outlook report* with permission from the Nature Publishing Group.[3] Figure courtesy of Gary Giovino.

However, as recently as the beginning of the 20th century, forms of smokeless tobacco (tobacco products that are chewed, sucked or sniffed) were the dominant form of tobacco use (Fig 8.1). The earliest of these products, a dry snuff typically used nasally, is now rarely used even in countries where it is still freely available (including Germany, India, USA and some parts of Africa). In some countries, however, including India and Sweden, other smokeless tobacco products continue to compete with cigarettes as the most widely used products.[4,5] The use of smokeless tobacco in its various forms is, in fact, common throughout the world: chewing tobacco and oral snuff (tobacco that is sucked rather than chewed) are used in North America; snus (a moist oral snuff) is widely used in Sweden and to a lesser extent Norway; and paan and gutka (both oral tobacco products) are used in South East Asia.

All of these varieties deliver pharmacologically active doses of nicotine by direct absorption through the lining of the mouth. These different products vary as much as 130-fold in their content and delivery of tobacco toxins.[6] Some have very high concentrations of tobacco-specific nitrosamines (TSNAs) and are a significant cause of oral cancer,[7] while others, including the snus used in Sweden, have relatively low concentrations of TSNAs and appear either to not cause cancer, or to present a much lower level of risk.[8–10] Given the harmful effects of nicotine on the fetus, all of these products are potentially harmful in pregnancy. It has been suggested that the variety used in Sweden (snus) is around 90% less harmful to health than smoking,[11] and has had a net beneficial effect on the health of men in Sweden by acting to reduce the number of daily smokers.[4]

Awareness that smokeless tobacco can compete with cigarettes for market share, and that some forms of smokeless tobacco may be markedly less harmful to health than smoked tobacco, has led to a renewed interest within the public health community in the role of smokeless tobacco in reducing mortality and morbidity caused by smoking. This issue has caught the attention of some of the major cigarette manufacturers, who have begun to test-market new smokeless tobacco products around the world. However, before forming a reaction to such developments, it is important to understand the nature of these different types of smokeless tobaccos and their associated health risks. This chapter will describe the most commonly used forms of smokeless tobacco, the toxin constituents and nicotine deliveries of smokeless tobacco products, and summarise the epidemiological evidence on the known health effects of different forms of smokeless tobacco.

8.2 Common types of smokeless tobacco used around the world

There is a diverse range of smokeless tobacco products in use in different parts of the world, and a comprehensive review of all individual products is beyond the scope of this report. A useful summary guide, with photographs, is available from the US National Cancer Institute and Centers for Disease Control,[12] and online at: www.cancercontrol.cancer.gov/tcrb/stfact_sheet_combined10-23-02.pdf

8.2.1 South East Asia: paan, gutka and other products

Paan is a South Asian snack, which consists of mineral or slaked lime (calcium hydroxide), spices, sweeteners and the areca nut (which is sometimes inaccurately called 'betel nut') wrapped in a triangular package using betel leaves, and often held together with a toothpick. It is intended to be chewed. In India and other parts of South East Asia (Pakistan, Vietnam, Sri Lanka), tobacco is often added. In other countries (Taiwan, for example) the product is consumed without tobacco. The lime acts to keep the main psychoactive ingredients in the free-base or alkaline form, thus enhancing buccal absorption into the bloodstream. The areca nut contains the alkaloid arecoline, which promotes salivation (the saliva is stained red) and is itself a psychoactive stimulant with effects similar to those of nicotine.[13] This combination, known as a 'betel quid', has been used for several thousand years in many South Asian countries, often in association with important cultural ceremonies. Areca nut is also available combined with spices in ready-to-eat commercially produced pouches called paan masala. When tobacco is included, the product is called gutka.

Several other types of smokeless tobacco products commonly used in South East Asia include mishri, a pyrolysed tobacco product often used as dentifrice by women; zarda, a moist or dry chewing tobacco mixed with colourings, spices and perfumes, dried whole and chopped tobacco leaves; and nass, which comprises tobacco mixed with lime, ash and cotton oil. Many other forms are also available.

A representative, cross-sectional national family health survey of 301,984 adults aged 18 and older, from 92,447 households in 26 Indian states in 1998–9, found that 18% of Indian adults smoke tobacco and 21% chew tobacco.[14] Smokeless tobacco use was found to be more common than smoking among women (13% and 3% respectively), whereas smoking was more common than chewing in men (33% and 29% respectively). Previous studies in India showed a broadly similar pattern.[15] This suggests that, with an adult population of over 760 million people, there are over 160 million users of smokeless tobacco in India alone (114 million men and 48 million women).

8.2.2 North America: moist snuff, dry snuff and chewing tobacco

The two main types of smokeless tobacco used in North America are moist snuff and chewing tobacco. North American moist snuff contains tobacco with relatively high concentrations of nitrates and is allowed to ferment both before and after packaging, resulting in the formation of carcinogenic tobacco-specific nitrosamines (TSNAs). American moist snuff also tends to be made from dark fire-cured tobacco, resulting in raised concentrations of the carcinogen benzo(a)pyrene. American snuff is sold in cans either as loose tobacco varying in consistency from that similar to tea-leaves to ground coffee, or as 'portion packed' in sachets like small tea-bags. Moist snuff is typically used by placing a loose or packaged portion of tobacco in the space between the gum and the cheek or lip. It is not usually chewed.

Chewing tobacco is sold in a variety of forms (primarily 'plug' or 'looseleaf'), is typically combined with syrup or other sweeteners and has a more dense texture allowing it to be chewed.

Dry snuff is made typically from fire-cured tobacco that has been fermented and ground into a powder. It was originally intended for nasal use, but women in the southern states of America have used dry snuff as an oral tobacco since the early 1800s.[16] Dry snuff is the oldest form of tobacco used in Europe. Winn et al studied elderly (mean age 69) female participants who used dry snuff orally.[17] Beyond this study, epidemiological evidence on the effects of dry snuff is very sparse.[18,19]

Use of smokeless tobacco by adults in the United States has now declined to around 5% of men and has remained at less than 1% of women. Use in the previous month by high school seniors has also declined from 20% in the mid-1980s to around 14% in 2003.[20] Within this general downward trend, however, use of moist snuff has increased by 66% over the past 15 years.[21,22]

8.2.3 Scandinavia: snus

The moist snuff traditionally used in Sweden is called 'snus'. The tobacco is air- and sun-cured, ground and processed by steam treatment (a process similar to pasteurisation) rather than fermentation. It is believed that the heat treatment prevents microbial fermentation and consequent nitrosamine formation.[4] Selection of air-cured tobaccos also minimises the content of polycyclic aromatic hydrocarbons (for example, benzo(a)pyrene). Snus is sold either loose or portion-packed, and the tins are typically kept refrigerated in stores. The product is placed between the gum and upper lip and does not typically result in increased salivation requiring spitting (as is common with North American 'spit tobacco'). The largest manufacturer of Swedish snus has introduced a manufacturing standard for its snus products, called the 'Gothiatek Standard', which specifies maximum permissible limits for toxins in the product (see Table 8.1).[4] The prevalence of daily snus use in Sweden has increased between 1976 and 2004, from approximately 9% to 22% in men and from 0% to 4% in women.

Table 8.1. The Gothiatek Standard, a voluntary market-based toxicity standard used for snus products by Swedish Match Tobacco Company.

Toxin	Maximum permissible limit
Nitrate	3.5 mg/kg
Tobacco-specific nitrosamines (TSNA)	5 mg/kg
N-Nitrosodimethylamine (NDMA)	5 µg/kg
Benzo(a)pyrene (BaP)	10 µg/kg
Cadmium	0.5 mg/kg
Lead	1.0 mg/kg
Arsenic	0.25 mg/kg
Nickel	2.25 mg/kg
Chromium	1.5 mg/kg

µg = microgram or 10^{-6}g. mg/kg ~ parts per million (ppm); µg/kg is equivalent to parts per billion (ppb). Limits based on 50% water content – double the limits for dry weight equivalents.

8.2.4 Sudan: toombak

Loose snuff, known locally as 'toombak', was introduced in Sudan over 400 years ago.[7] Toombak is produced from the *Nicotiana rustica* species of tobacco, and the fermented ground powder is mixed with an aqueous solution of sodium bicarbonate. The resultant product is moist with a strong aroma, and its use is widespread in Sudan, particularly among men (around 35% of men compared with 3% of women).

8.2.5 New smokeless tobacco products

Particularly in the past 15 years or so, various novel smokeless tobacco products have been developed and marketed. One example is a compressed nicotine lozenge made from powdered tobacco that dissolves in the mouth (marketed as Ariva). The manufacturer claims that it is low in tobacco toxins such as TSNAs. This product has been available in US pharmacies for several years but has not sold in significant volume. Other new products include a dissolvable, low-TSNA preparation called Stonewall Hard Snuff. In the United Kingdom a variety of unlicensed tobacco/nicotine lozenges (including Stoppers and Stubit) have been marketed through pharmacies as unlicensed smoking cessation aids.[23]

A number of new brands of smokeless tobacco claiming to differ from traditional brands have recently been launched in the United States. Revel, for example, is a portion-packed snuff tobacco that appears to be targeted at cigarette smokers for use when they cannot smoke. Other products, including Camel Snus and Taboka are also being test-marketed, but have not to date gained significant market share.

8.3 Potentially harmful constituents

Aside from nicotine, the main constituents of smokeless tobacco suspected of harming health are those included in the Gothiatek Standard (see Table 8.1). Smokeless tobacco is capable of delivering sufficient quantities of nicotine with sufficient speed to have reinforcing psychoactive effects and, as discussed in Chapter 5, is potentially dependence-forming in many users. Fig 8.2 shows the blood nicotine profile of one brand of Swedish snus in comparison with cigarettes and nicotine replacement therapies.[4]

Of the other constituent toxins, most attention has focused on four tobacco-specific nitrosamines (TSNAs): N-Nitrosonornicotine (NNN), N-Nitrosoanatabine (NAT), N-nitrosoanabasine (NAB), and 4(methylnitrosamino)-1-(3-pyridyl)-1-butanone (NNK). The International Agency for Research on Cancer (IARC) has concluded that NNN and NNK are carcinogenic to humans,[24] and some studies have focused on measuring concentrations of these TSNAs. However, other toxins, such as the known carcinogen benzo(a)pyrene, are also present in some forms of smokeless tobacco and could potentially cause harmful effects.

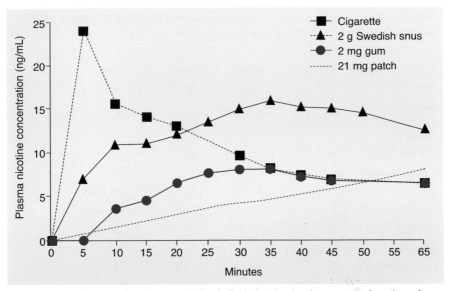

Fig 8.2 Venous blood concentrations (ng/ml) of nicotine in plasma, as a function of time for various nicotine delivery systems. All plasma nicotine concentrations have been reconfigured such that the pre-absorption level starts at 0 ng/ml (to take out baseline differences). Reproduced from Foulds *et al* with permission from the BMJ Publishing Group.[4]

The types of toxins measured and the method of reporting toxin concentrations have varied across studies, sometimes making comparisons difficult. Table 8.2 summarises the results of studies of concentrations of TSNAs in moist snuff products, conducted from 1990 to 2000. Note that these analyses are based on dry weight. Much higher concentrations of NNN and NNK are found in Sudanese toombak than North American smokeless tobacco, which in turn has higher concentrations than Swedish snus (Table 8.2).

Since 2000, a number of studies have compared toxin content in a variety of smokeless products. Osterdahl *et al* reported on analyses of samples of Swedish snus from eight manufacturers (including seven small-scale manufacturers).[29] They reported that the mean level of the total TSNA contained in moist snus was around 1.0 μg/g (based on wet weight), and suggested that the level of TSNAs in moist snuff on the Swedish market has been reduced greatly since the middle of the 1980s.

Rodu and Jansson reported on TSNA levels in a variety of tobacco products available in 2003.[30] The main results of this study show negligible TSNA content in Ariva, very low TSNA levels in Revel, Exalt, Swedish snus and North American chewing tobacco, moderate levels in US moist snuff, and very high levels in some brands of US dry snuff (Table 8.3).[30]

Table 8.2. Mean tobacco-specific nitrosamine content (and range) of moist snuff products from various sources based on dry weight.

Country and brand (year sampled)	Manufacturer	Nicotine (mg/g)	NNK (µg/g)	NNN (µg/g)	Total TSNA (µg/g)
Sweden					
Three brands* (1990–91)[25]	Swed. Match		1.4–2.1	5.2–5.7	9.2–11.2
Ettan Snus (2000)[26]	Swed. Match		0.5	1.1	2.8
Sudan (toombak)[27]					
5 samples* (1990)		32.2–102.4	630–7,870	830–3,805	
5 samples* (1990)		8.4–26.0	1,140–2,790	420–1,550	
3 samples* (1993)			188–362	241–369	
United States					
2 samples* (1991)[25]		18.6–20.6	0.5–0.8	4.8–8.0	
1 sample* (1992)[25]		16.7	0.6	5.6	
Copenhagen (1994)[28]	USSTC	12 (11.3–12.7)	1.9 (1.3–2.5)	8.7 (7.3–10.1)	17.2 (14.2–20.2)
Skoal, Original fine cut (1994)[28]	USSTC	11.9 (10.7–13.4)	1.3 (1.2–1.4)	8.2 (6.9–9.5)	14.9 (12.4–17.4)
Skoal Bandts Straight (1994)[28]	USSTC	10.1 (9.3–10.9)	0.9 (0.6–1.2)	5.1 (4.1–6.1)	8.2 (6.5–9.9)
Kodiak Wintergreen (1994)[28]	Conwood	10.9 (10.1–11.7)	0.6 (0.4–0.8)	6.3 (5.2–7.4)	11.0 (8.6–13.4)
Hawken Wintergreen (1994)[28]	Conwood	3.2 (3.0–3.4)	0.2 (.16–.24)	3.1 (2.8–3.4)	4.1 (3.7–4.5)
Skoal (2000)[26]	USSTC		4.3	20.8	64.0
Copenhagen(2000)[26]	USSTC		3.4	14.3	41.1
Timber Wolf(2000)[26]	Swed. Match		0.95	3.0	7.5
Silver Creek(2000)[26]	Swisher		17.8	41.4	127.9

NNK = 4(Methylnitrosamino)-1-(3-pyridyl)-1- butanone; NNN = N-Nitrosonornicotine; TSNA = tobacco-specific nitrosamines.
* Brand not specified in study.

Stepanov *et al* reported on the concentrations of TSNAs in smokeless tobacco and related products sold in India,[31] and also in new smokeless products sold in the United States and Sweden (2006).[32] The main results (based on wet weight) demonstrate a wide variation in TSNA concentrations across products, from <0.05 µg/g NNN in Ariva, and Stonewall hard tobaccos, to >35 µ/g NNN in two brands of Indian Khani (Table 8.4).[31,32] There was some overlap in TSNA concentrations found in Swedish snus and different brands of traditional American smokeless products, but the new Revel brand had lower concentrations than both of those

Table 8.3. TSNA levels[a] in American and Swedish tobacco products, 2003.

Tobacco type/ brand name	Dry matter (%)	NNN	NNK	NAT	NAB	Total TSNAs
Cigarettes						
Camel	91	3.4	0.8	2.2	0.1	6.5
Marlboro	91	3.5	1.5	1.9	0.1	7.0
Chewing tobacco						
Beech Nut	78	3.0	0.8	0.8	0.1	4.7
Red Man	76	1.0	0.3	0.5	0.0	1.8
Oliver Twist – Tropical	81	0.9	0.1	0.5	0.0	1.5
Oliver Twist – Senior	80	1.7	0.3	1.3	0.1	3.4
Moist snuff, US						
Skoal Straight Long Cut	46	5.2	1.6	3.8	0.3	10.9
Skoal Bandits Straight	51	4.2	0.7	1.8	0.1	6.8
Skoal Wintergreen	44	2.7	0.6	1.4	0.1	4.8
Copenhagen	46	5.5	1.3	5.0	0.3	12.1
Copenhagen pouches	46	2.4	0.4	1.5	0.1	4.5
Hawken Wintergreen	74	4.8	1.1	1.1	0.3	7.3
Kodiak Wintergreen	48	6.4	0.7	4.8	0.4	12.3
Moist snuff, Sweden						
General	45	1.1	0.4	0.6	0.1	2.1
Ettan	47	1.1	0.3	0.6	0.1	2.0
Catch Licorice	52	1.0	0.4	0.6	0.0	2.0
Goteborgs Rape	44	1.1	0.4	0.6	0.0	2.2
Grovsnus	45	1.1	0.5	0.6	0.1	2.2
Dry snuff						
Bruton	94	287	922	77	31	1,219
Red Seal	94	210	280	210	32	1,096
Dental Sweet	93	19	6.5	14	1.2	41
Scotch	93	21	22	20	2.1	65
New products						
Revel	95	1.3	0.2	0.7	0.1	2.3
Ariva	97	0.0	<0.1	0.0	0.0	<0.1
Exalt	91	3.1	1.1	1.5	0.2	5.8

[a] All concentrations in parts per million based on dry weight.
NNN = N-Nitrosonornicotine; NAT = N-Nitrosoanatabine; NAB = N-Nitrosoanabasine; NNK = 4(Methylnitrosamino)-1-(3-pyridyl) – butanone; TSNA = tobacco-specific nitrosamines.
Data from Rodu et al.[30] Reproduced with permission from Critical Reviews in Oral Biology and Medicine.

Table 8.4. Tobacco-specific nitrosamines, nitrate, nitrite and nicotine in Indian, US and Swedish smokeless tobacco and related products.

Product	Tobacco-specific nitrosamines (µg/g)					NO_3 (µg/g)	NO_2 (µg/g)	Nicotine (mg/g)
	NNN	NAT	NAB	NNK				
Khani								
Raja	76.9	13.8	8.83	28.4		705	1,020	21.3
Hans Chhap	39.3	4.83	3.78	2.34		1,090	1,410	19.6
Zarda								
Goa 1000	8.36	1.98	0.48	3.09		966	2.20	14.6
Moolchand Super	6.47	0.64	0.46	1.64		1,320	ND	15.0
Sanket 999	7.77	1.51	0.36	1.99		1,910	2.08	65.0
Baba 120	4.81	1.40	0.19	1.07		1,700	1.63	44.2
Shimla	19.9	1.53	1.19	2.61		1,360	2.53	13.8
Other tobacco								
Hathi Chhap	2.75	1.53	0.23	0.85		2,760	1.97	39.5
Gai Chhap	19.2	11.9	1.57	2.61		2,950	8.40	47.8
Miraj	1.74	0.35	0.12	0.08		1,420	13.6	15.6
Mishri								
Shahin	4.21	2.55	0.15	0.87		1,720	5.18	21.0
Gutka								
Star 555	0.47	0.07	0.02	0.13		417	1.61	6.77
Manikchand	0.38	0.05	0.01	0.12		43.9	2.00	3.22
Zee	0.32	0.05	0.01	0.08		62.3	3.42	3.31
Tulsi Mix	0.69	0.07	0.02	0.31		184	2.58	5.67

continued

Table 8.4. Tobacco-specific nitrosamines, nitrate, nitrite and nicotine in Indian, US and Swedish smokeless tobacco and related products – continued.

Product	Tobacco-specific nitrosamines (µg/g)					NO_3^- (µg/g)	NO_2^- (µg/g)	Nicotine (mg/g)
	NNN	NAT	NAB	NNK				
Gutka – continued								
Wiz	0.31	0.04	0.02	0.13		215	2.82	1.67
Kuber	0.32	0.03	0.01	0.13		47.3	4.50	1.23
Pan Parag	0.44	0.06	0.02	0.12		332	2.84	2.67
Zatpat	1.09	0.08	0.05	0.43		171	1.99	5.48
Vimal	0.09	0.01	ND	0.04		268	1.58	6.82
Josh	0.49	0.08	0.03	0.20		252	1.74	11.4
Supari								
Goa	ND	ND	ND	ND		7.5	4.71	NA
Moolchand	ND	ND	ND	ND		8.5	2.48	NA
Rajanigandha	ND	ND	ND	ND		8.8	3.34	NA
Sanket	ND	ND	ND	ND		8.5	4.27	NA
Shimla	ND	ND	ND	ND		8.0	6.56	NA
Creamy snuff/toothpaste								
IPCO	3.32	0.53	0.11	1.31		580	ND	4.71
Dentobac	2.52	1.49	0.07	2.16		232	ND	7.71
Snuff								
Click	0.56	0.38	0.02	0.24		2.260	ND	71.4

continued

Table 8.4. Tobacco-specific nitrosamines, nitrate, nitrite and nicotine in Indian, US and Swedish smokeless tobacco and related products – continued.

Product	NNN	NAT	NAB	NNK	NO$_3^-$ (µg/g)	NO$_2^-$ (µg/g)	Nicotine (mg/g)
Tobacco-specific nitrosamines (µg/g)							
Tooth powder							
Baidyanath	0.04	ND	ND	ND	48.6	ND	0.72
New Roshanjyot	ND	ND	ND	ND	11.6	1.25	0.25
Dabur	0.04	ND	ND	ND	27.6	ND	0.58
Reference snuff							
Kentucky IS3	3.39	3.15	0.25	0.94	3.86	6.35	36.2
New tobacco products							
Hard snuff							
Ariva	0.019	0.12	0.008	0.037			
Stonewall	0.056	0.17	0.007	0.043			
Swedish snus							
General	0.98	0.79	0.06	0.18			
Exalt							
Purchased in Sweden	2.3	0.98	0.13	0.27			
Purchased in the US	2.1	0.68	0.05	0.24			
Revel							
Mint flavored	0.62	0.32	0.018	0.033			
Wintergreen flavored	0.64	0.31	0.017	0.032			

continued

Table 8.4. Tobacco-specific nitrosamines, nitrate, nitrite and nicotine in Indian, US and Swedish smokeless tobacco and related products – continued.

Product	Tobacco-specific nitrosamines (µg/g)				NO_3 (µg/g)	NO_2 (µg/g)	Nicotine (mg/g)
	NNN	NAT	NAB	NNK			
Tobacco-free snuff							
Smokey Mountain	ND	ND	ND	ND			
Nicotine replacement therapy products							
NicoDerm CQ (patch, 4-mg nicotine)[f]	ND	ND	ND	0.008			
Nicorete (gum, 4-mg nicotine)[f]	0.002	ND	ND	ND			
Commit (lozenge, 2-mg nicotine)[f]	ND	ND	ND	ND			
Conventional tobacco products							
US snuff							
Copenhagen							
Snuff	2.2	1.8	0.12	0.75			
Long Cut	3.9	1.9	0.13	1.6			
Skoal							
Long Cut Straight	4.5	4.1	0.22	0.47			
Bandits	0.9	0.24	0.014	0.17			
Kodiak							
Ice	2.0	0.72	0.063	0.29			
Wintergreen	2.2	1.8	0.15	0.41#			

ND = none detected; NNN = N-Nitrosonornicotine; NAT = N-Nitrosoanatabine; NAB = N-Nitrosoanabasine; NNK = 4(Methylnitrosamino)-1-(3-pyridyl) – butanone. Data from Stepanov et al.[31,32]

categories (though slightly higher than the new 'hard tobacco' brands). All the TSNA levels found in smokeless tobacco were higher than those found in nicotine replacement products, which contained only trace levels of TSNAs.

McNeill *et al* reported on a variety of toxin concentrations (based on dry weight) in Asian smokeless tobacco products purchased in the UK, as well as select products from other countries (Table 8.5).[6] This study found Ariva (hard tobacco) to have undetectable levels of TSNAs, and over 100-fold variations in toxin concentrations across other products. One Asian zarda product (Hakim Pury) had particularly high TSNA levels, and the leading US snuff product (Copenhagen) had notably higher concentrations of benzo(a)pyrene.

The implications of the data from these various sources are that there are very large differences in toxin concentrations between different smokeless tobacco products, and that development and production of new smokeless products with very low toxin concentrations is commercially feasible. It is therefore possible that some of these products might offer substantially less harmful alternatives to smoked tobacco or to some existing forms of smokeless tobacco.

8.4 Overview of health effects of smokeless tobacco products

Over the past decade, studies have examined the effects of paan/areca nut use with and without tobacco added, and have confirmed that the paan use without tobacco is highly carcinogenic.[33–36] Merchant *et al* found that people using paan without tobacco were 9.9 times more likely to develop oral cancer and those using paan with tobacco were 8.4 times more likely to develop oral cancer, compared with non-users after adjustment for other covariates.[33] This study and others identified an independent effect of paan without tobacco in the causation of oral cancer, and a role of areca-specific nitrosamine as a causative carcinogen has been suggested.[37] When discussing the potentially harmful constituents of smokeless tobacco products it is therefore important to distinguish the constituents intrinsic to the smokeless tobacco, as opposed to the constituents that are sometimes added to or taken along with tobacco. The most important example of the latter category is areca nut. Areca nut is clearly carcinogenic when consumed without tobacco,[36] and so studies purporting to provide evidence on the effects of smokeless tobacco that include or fail to distinguish those users also using areca nut (or smoked tobacco) cannot provide clear evidence on the role of smokeless tobacco *per se*. Given that areca nut also affects cardiovascular function (albeit in a manner that has been insufficiently studied),[13] this point also applies to studies of the effects of smokeless tobacco on heart disease that do not distinguish users who mix tobacco with areca nut.

Table 8.5. Content of smokeless tobacco products.[a]

Brand	Moisture (% w/w)	TSNA (µg/g)	BaP (ng/g)	NDMA (ng/g)	Nitrite (µg/g)	Chromium (mg/kg)	Nickel (mg/kg)	Arsenic (mg/kg)	Lead (mg/kg)	Nicotine (mg/g)	Average pH	Free nicotine (mg/g)
UK purchased products												
Gutkha products												
Manikchard	1.68	0.289	0.40	ND	ND	0.26	1.22	0.04	0.15	3.1	9.19	3.0
Tulsi Mix	1.25	1.436	1.28	ND	ND	0.33	1.43	0.06	0.19	8.2	9.52	8.0
Zarda products												
Hakim Pury	4.91	29.705	0.32	ND	ND	2.15	5.35	0.29	1.36	42.7	6.00	0.4
Dalal Misti Zarda	8.96	1.574	8.89	ND	ND	0.87	2.09	0.11	1.14	8.6	6.15	0.1
Baba Zarda (GP)	7.88	0.716	2.04	ND	ND	2.34	5.88	0.24	1.18	48.4	5.32	0.1
Tooth cleaning powder												
A. Quardir Gull	3.35	5.117	5.98	7	ND	3.56	5.31	0.46	1.39	64.0	9.94	63.2
Dried tobacco leaves												
Tobacco Leaf	5.16	0.223	0.11	ND	ND	2.34	4.37	0.20	1.06	83.5	5.52	0.3

continued

Table 8.5. Content of smokeless tobacco products[a] – continued.

Brand	Moisture (% w/w)	TSNA (µg/g)	BaP (ng/g)	NDMA (ng/g)	Nitrite (µg/g)	Chromium (mg/kg)	Nickel (mg/kg)	Arsenic (mg/kg)	Lead (mg/kg)	Nicotine (mg/g)	Average pH	Free nicotine (mg/g)
Products purchased outside UK												
Baba 120 (India)	13.18	2.361	2.83	ND	ND	2.08	2.94	0.40	1.56	55.0	4.88	0.04
Snus (Sweden)	45.84	0.478	1.99	ND	ND	1.54	2.59	0.30	0.50	15.2	7.86	6.3
Ariva (US)	2.40	ND	0.40	ND	ND	1.40	2.19	0.12	0.28	9.2	7.57	2.4
Copenhagen (US)	48.10	3.509	19.33	ND	6.7	1.69	2.64	0.23	0.45	25.8	7.39	4.9
Detection Limits		0.025 for each		5	0.2							

[a] All figures are averages of two measurements, except for pH which gives the average of three measurements. On average, measurements agreed by less than 10%. Total TSNA = total tobacco-specific N-nitrosamines (NNK+NNN+NAB); BaP = Benzo(a)pyrene; NDMA = N-nitrosodimethylamine. Data from McNeill et al.[6]

Over recent years, a number of reviews of health effects of smokeless tobacco or specific types of smokeless tobacco have been published. A systematic review by Critchley and Unal concluded that chewing betel quid (including areca nut) with tobacco is associated with a substantial risk of oral cancer, but that most studies based in the United States or Scandinavia did not find statistically significant effects, though moderate positive associations could not be ruled out as many studies were too small to be conclusive.[38] Critchley and Unal also concluded that there may be an association between smokeless tobacco use and cardiovascular disease, but that further studies with adequate sample sizes are required.[39] The rest of this section will examine the evidence for an association between smokeless tobacco and specific types of disease.

8.4.1 Oral cancer

One of the biggest concerns about the use of smokeless tobacco stems from the large body of evidence from a number of countries showing that people who use smokeless tobacco have a higher risk of developing cancer of the mouth. A 2001 US Institute of Medicine (IOM) report stated that, 'A large number of studies in India, including cohort, case-control, and intervention studies, support an association between oral cancer and smokeless tobacco, and these studies are consistent, strong, coherent and temporally plausible.'[40] The IOM report stated that toombak users in Sudan also have a much higher risk of oral cancer than non-users and that, 'In spite of conflicting US data, it can be concluded that snuff use in the United States also increases the risk of oropharyngeal cancers.'

Some of the best evidence for a carcinogenic effect of smokeless tobacco in the United States comes from a study by Winn and colleagues.[17] This case-control study focused on women in North Carolina, USA and found that white never smoking women who used oral dry snuff powder had a relative risk of over 4.2 (2.6–6.7) for developing oral and pharyngeal cancer. Women who had used dry snuff for 50 years had a 50-fold increase in risk for some oral cancers. It should be noted that only a tiny minority of smokeless users use this type of tobacco in the United States. The data in Table 8.3 also suggest that some forms of dry snuff have much higher concentrations of carcinogens than any other smokeless products.

It is difficult to separate out the effects of areca nut, smoking and smokeless tobacco in South East Asian studies as these habits are typically highly correlated. Since the publication of the IOM report, additional large studies of the relationship between smokeless tobacco use and oral cancer have been published. In one of the largest studies, Henley *et al* examined the association between exclusive use of oral tobacco (compared with never tobacco users) and mortality from various causes in the Cancer Prevention Studies (CPS-I and II) conducted by the American Cancer Society.[41] The CPS-I analysis included 7,745 exclusive smokeless tobacco users (median age 62) and almost 70,000 never tobacco users recruited in 1959 and followed up to determine cause of death by 1971. CPS-II analyses included 2,488 smokeless users (median age 57) and over 110,000 never tobacco users recruited in 1982 and followed up to determine cause of death by 2000. The studies both ascertained tobacco use at the outset of the study and assumed that use did not change throughout the follow-up period (Table 8.6).[41]

Neither analysis found evidence of a statistically significant increase in the risk of death from oral cancer. The adjusted hazard ratio estimate in CPS-I was 2.02 and 0.9 in CPS-II. The confidence intervals on both of these estimates were wide. The CPS-I estimate is derived from a larger number of smokeless tobacco users but with shorter follow-up and included only 13 deaths from oral cancer. The CPS-II population of 2,488 smokeless users with a median age of 57 at enrolment and followed for 18 years generated only one death from oral cancer in exclusive smokeless tobacco users and none in former users. Together, these findings suggest that any effect, if present, is very small. Five-year survival from oral cancer has been around 50% for the past 30 years,[42] and so it is unlikely that the lack of effect was due to a large number of incident cases occurring but not causing death.

Accortt et al analysed the National Health and Nutrition Examination Survey (NHANES-1) in the United States, comparing the incidence of oral cancer in 414 smokeless tobacco users and 2,979 never tobacco users.[43] The participants were aged 45–75 at enrolment in the mid-1970s and were followed up around 10 years later. No cases of oral cancer were observed among the smokeless users, and it was concluded that the standardised incidence rate was not increased among smokeless users.

Rosenquist et al conducted a case-control study in southern Sweden to examine the risk of oral cancer in snus users, and found that while both alcohol consumption (odds ratio = 2.6) and smoked tobacco consumption (odds ratio = 4.7) were associated with oral cancer, there was no increased risk of oral cancer among snus users in southern Sweden (odds ratio 1.1).[9]

Roosaar et al conducted a record-linkage 20-year follow-up study of 267 Swedish snus users who presented originally with snus-related oral lesions.[8] They found three cases of oral cancer, none of which was at the original lesion site (that is, none was at the spot where the participant originally placed his smokeless tobacco), and the original snus-induced lesions had disappeared among the 62 individuals who had quit using snus. They concluded that the incidence of oral cancer in this cohort of individuals with snus-induced lesions was slightly higher than expected, but that cancers rarely occur at the site of lesions observed in the distant past.

Table 8.6. Adjusted mortality hazard ratios (and 95% confidence intervals) associated with the use of smokeless tobacco, relative to never tobacco use among men who never used other tobacco products, CPS-I (1959–1972) and CPS-II (1982–2000), in the United States.

Cause of death (ICD-7 code)	Multivariate-adjusted hazard ratio, CPS-I	Multivariate-adjusted hazard ratio, CPS-II
All causes	1.17 (1.11–1.23)	1.18 (1.08–1.29)
All cancers (140–239)	1.07 (0.95–1.20)	1.19 (1.02–1.40)
Oropharynx cancer (140–148)	2.02 (0.53–7.74)	0.90 (0.12–6.71)
Digestive system cancer (150–159)	1.26 (1.05–1.52)	1.04 (0.77–1.38)
Lung cancer (162–163)	1.08 (0.64–1.83)	2.00 (1.23–3.24
Genitourinary system cancer (177–181)	0.97 (0.77–1.22)	1.15 (0.85–1.56)
Other cancers (160–161, 163–175, 190–205)	0.90 (0.71–1.14)	1.49 (1.04–2.14)
Hematopoietic cancers (200–208)		0.95 (0.60–1.51)
Cardiovascular disease (330–468)	1.18 (1.11–1.26)	1.23 (1.09–1.39)
Coronary heart disease (420)	1.12 (1.03–1.21)	1.26 (1.08–1.47)
Cerebrovascular disease (330–334)	1.46 (1.31–1.64)	1.40 (1.10–1.79)
Other cardiovascular (335–398, 400–419, 421–468)	1.05 (0.91–1.22)	1.07 (0.82–1.39)
Other causes (001–138, 206–289, 470–E999)	1.17 (1.06–1.30)	1.11 (0.97–1.25)
Diabetes (260)	0.88 (0.53–1.47)	1.12 (0.55–2.29)
Respiratory system diseases (470–527)	1.28 (1.03–1.59)	1.11 (0.84–1.45)
Influenza, pneumonia (480–493)	1.16 (0.88–1.51)	0.85 (0.56–1.29)
Chronic obstructive pulmonary disease (480–493)	1.86 (1.12–3.06)	1.28 (0.71–2.32)
Digestive system disease (530–587)	1.49 (1.14–1.93)	1.38 (0.92–2.07)
Colitis and other intestinal (570–578)	1.42 (0.94–2.12)	1.12 (0.65–1.92)
Cirrhosis (581)	1.49 (0.87–2.56)	3.02 (1.60–5.69)
Genitourinary system diseases (590–637)	1.34 (1.00–1.80)	1.02 (0.62–1.69)
Nephritis and other kidney disease (590–603)	1.37 (0.98–1.92)	1.01 (0.53–1.93)
External causes (E800–E999)	1.05 (0.84–1.32)	1.26 (0.93–1.70)

CPS-I and CPS-II = Cancer Prevention Studies conducted by the American Cancer Society; ICD = World Health Organization International Classification of Diseases.
Data from Henley et al.[41] Reproduced with kind permission from Springer Science and Business Media.

Rodu and colleagues reviewed the relationship between oral cancer and smokeless tobacco.[30] They found no evidence of increased risk of oral cancer from chewing tobacco or most types of moist snuff typically available in North America, but did find a significantly increased risk of oral cancer from use of some types of dry snuff (Fig 8.3). Most recently, Luo *et al* reported on the risk for oral cancer among a cohort of almost 280,000 Swedish male construction workers followed up for 20 years.[44] As in the studies mentioned above,[4,8,9] they found that snus use did not increase risks of oral cancer (rates were in fact slightly lower in snus users than in non-tobacco users), whereas smoking more than doubled the risk (Fig 8.4).[44]

Overall, therefore, it seems clear that some forms of smokeless tobacco, primarily those with the highest concentrations of carcinogens, cause oral cancer. However, it is also clear that the risk of oral cancer associated with use of low-TSNA

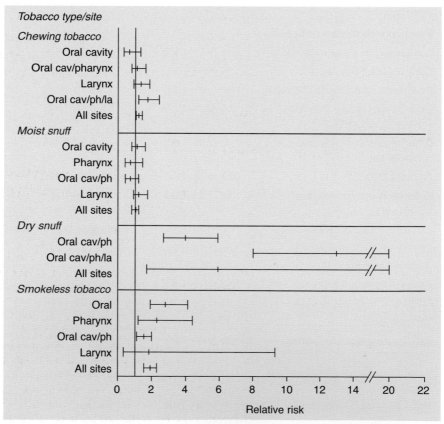

Fig 8.3 Summary relative risks (where 1 = identical risk to never tobacco user) for oral cancer and related sites according to SLT product type. Oral Cav = Oral Cavity; Ph = Pharynx; La = Larynx. Adapted from Rodu *et al.*[30] Reproduced with permission from *Critical Reviews in Oral Biology and Medicine*.

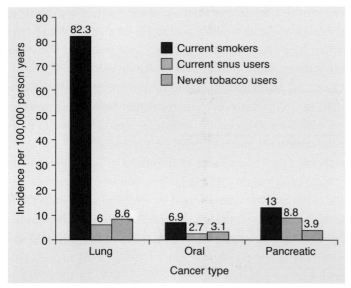

Fig 8.4 Incidence rate for three types of cancer (by 2004) as a function of tobacco use status in male Swedish construction workers at recruitment (1978–92). Adapted from data in Luo et al.[44]

tobacco products such as Swedish snus is small, and possibly non-existent. The main evidence associating smokeless tobacco with oral cancer comes predominantly from studies in populations combining smokeless tobacco with other toxins (such as areca nut) or from populations using products that contain higher concentrations of carcinogenic compounds than are present in current moist snuff or new smokeless tobacco products available in the United States or Scandinavia.

8.4.2 Pancreatic cancer

It has also been suggested that smokeless tobacco may increase risks of pancreatic cancer, and there are plausible mechanisms for such effects. Smoking is a known cause of pancreatic cancer and TSNAs have been identified in pancreatic secretions of smokers.[45] Recently, Stepanov and Hecht found higher levels of some TSNAs in the urine of smokeless tobacco users compared with smokers, so it is plausible that levels will pass through the pancreas in smokeless tobacco users.[46]

Bofetta et al reported a prospective cohort study of 10,136 Norwegian men who were enrolled in 1966 and followed to diagnosis by 2001, and of whom 32% used smokeless tobacco at some stage.[47] This study reported relative risks of eight types of cancer in current, former, and ever smokeless users. Overall, there was little consistent evidence of increased cancer risk among smokeless users. Among ever or current users, relative risks greater than one were reported for cancer at four different sites (oral/pharyngeal, oesophagus, stomach, pancreas) and four below one (lung cancer, lung adenocarcinoma, kidney and bladder). The effect on pancreatic cancer was significant in ever users (relative risk 1.67; 95% confidence interval 1.12 to 2.50) and borderline significant in current users (relative risk 1.60; 95% CI 1.00 to 2.55). This effect was restricted, however, to analyses including snus users who were also current or former smokers. Among non-smoking snus users the relative risk was 0.85 (95% CI 0.24 to 3.07).

A case-control study from the United States of the effects of smokeless tobacco use on risk for pancreatic cancer compared 154 lifelong non-smokers newly diagnosed with pancreatic cancer and 844 controls aged 30–79.[48] Estimated odds ratios for pancreatic cancer were all increased, though non-significantly, in users of cigars, pipes or smokeless tobacco. In ever users of smokeless tobacco the odds ratio was 1.4 (95% CI 0.5 to 3.6) and in current users it was 1.1 (95% CI 0.4 to 3.1). This effect was dose related, with an adjusted odds ratio of 3.5 (95% CI 1.1 to 11) in those defined as heavy users (more than 2.5 ounces of smokeless tobacco per week) compared with 0.3 (95% CI 0.04 to 2.5) in light users (less than 2.5 ounces per week).

The findings of the first of these studies have been challenged on numerous grounds, including that the observed effect of smokeless tobacco was restricted to current smokers, and that the analysis did not correct for potentially important confounding by alcohol intake.[47] However, the recent study by Luo et al also found a significantly increased rate of pancreatic cancer among snus users versus never tobacco users (about a doubling of risk), and both of these groups had lower risks than smokers.[44] The implication of these findings is that smokeless tobacco, including the snus consumed in Sweden over the past half century, appears to be associated with an increased risk of pancreatic cancer.

8.4.3 Lung cancer

In the CPS-II study described above there was a significantly increased risk of death from lung cancer in smokeless tobacco users (hazard ratio 2.0; 95% CI 1.23 to 3.24).[41] This effect was not seen in the CPS-I study (hazard ratio 1.08; 95% CI 0.64 to 1.83), and was surprising since the mechanism whereby use of smokeless tobacco in the mouth may cause cancer of the lung is not clear. It has been suggested that increased systemic exposure to NNK, a powerful systemic lung carcinogen, could be involved,[49] though other explanations for the finding, particularly bias arising from increased passive smoke exposure among smokeless tobacco users, have also been suggested.[50] This latter interpretation is supported by the fact that the raised age-adjusted hazard ratio for death from COPD, which is strongly related to smoke exposure, was 1.8 (95% CI 1.01 to 3.23) among smokeless users in CPS-II. To our knowledge, the observation of a raised risk of lung cancer in smokeless tobacco users has not been made in any other study, whereas a number have reported no increase in lung cancer risk among smokeless users.[44,47,51,52] In the Luo et al study, snus users had a slightly lower risk of lung cancer than never tobacco users, whereas smokers' risks were increased by approximately 10-fold.[44] Similarly, in the study by Bolinder et al, the relative risk for lung cancer among snus users compared with never tobacco users was 0.8 among men aged 55–65, whereas the relative risk was 30.6 (95% CI 14.6 to 64.1) for smokers of at least 15 cigarettes per day (again compared with never tobacco users).[51]

8.4.4 Smokeless tobacco and all cancer mortality

Bolinder and colleagues found a non-significant relative risk of death from cancer of 1.1 for snus users compared with never tobacco users (95% CI 0.9 to 1.4) in a prospective study of Swedish construction workers that included a relatively large sample, many of whom had used snus for over 40 years.[51] In the 1,734 snus users aged 55–65, most of whom would presumably have used snus for over 35 years, the relative risk for cancer death compared with non tobacco users was 1.0. This study found significantly increased all-cause mortality in snus users compared with never tobacco users, largely due to elevated cardiovascular mortality.

Overall, the numerous studies examining the effect of Swedish snus on cancer risk are fairly consistent in finding no increase in all-cause cancer among snus users.[51,53–55] Many of the Swedish studies of the relation between snus and cancer were robust enough to detect significant effects for tobacco smoking,[51,53,56] or alcohol use,[54,56] though these effects were generally much larger than those related to snus. Lagergren et al reported that combined smoking and alcohol use was strongly associated with oesophageal squamous cell carcinoma, with an odds ratio of over 23 for heavy users compared to never users.[56] The main exception has been the Luo study that found no increased risk of lung or oral cancer, but an increased risk of pancreatic cancer in snus users.[44] The relative magnitude of these risks is shown in Fig 8.4. It is particularly noteworthy that in the analyses by Luo et al of the whole male cohort, men who had ever used snus were significantly less likely to develop lung or oral cancer, and non-significantly less likely to develop pancreatic cancer, as compared with men who had never used snus (presumably because snus use reduced smoking).[44] Thus, it appears that, despite finding an association between snus use and pancreatic cancer, this study demonstrates that snus availability and use instead of cigarettes may have produced a large reduction in cancer incidence – primarily via reduced rates of lung cancer among smokers who switched. The majority of the snus users who developed pancreatic cancer in the study by Luo *et al* had used snus before the 1980s. Since that time, the level of carcinogens in snus has reduced.[29]

8.4.5 Smokeless tobacco and cardiovascular risk

Early studies of smokeless tobacco and cardiovascular diseases were primarily carried out in Sweden, and produced inconsistent results. Bolinder *et al* reported on a 12-year follow up of 135,000 Swedish construction workers, and found an increased risk of death from myocardial infarction in snus users compared with never tobacco users (relative risk 1.4; 96% CI 1.2 to 1.6).[51] Huhtasaari *et al* reported on two studies, neither of which found significantly increased cardiovascular risks for snus users, and one of which reported a significantly reduced risk of non-fatal myocardial infaction in snus users when compared with never regular tobacco users (odds ratio 0.58; 95% CI 0.35 to 0.94).[57,58] Each of these studies reported significantly increased risks for smokers, with larger effect sizes than those for snus users. Much of this evidence has been synthesised in previous reviews, which have concluded that, while most of the evidence was not supportive of a significant relationship, the inconsistencies in the study outcomes mean that a moderate effect of smokeless tobacco use on cardiovascular disease cannot be excluded.[4,39,58,59]

Since those reviews were published, five further studies have been reported.[41,60–63] Hergens *at al* reported a case-control study in 1,432 Swedish men who had a myocardial infarction in 1992–4, and 1,810 controls.[60] After adjustment for age, area and smoking, the relative risk for a first myocardial infarction was 1.1 (95% CI 0.8 to 1.5) for former snus users, and 1.0 (95% CI 0.8 to 1.3) for current snus users. The hypothesis that smokeless tobacco increases the risk of myocardial infarction was therefore not supported by these findings. Johansson *et al* conducted a 12-year follow-up study in 3,120 healthy Swedish men to establish risk factors for coronary heart disease.[61] The adjusted hazard ratio for never smoking snus users compared with never tobacco users was 1.41 (95% CI 0.61 to 3.28), whereas both smokers and former smokers had higher and significantly increased risks.

The recent analyses of the effects of smokeless use in the CPS-I and CPS-II studies found consistently raised hazard ratios for cardiovascular disease as a whole, coronary heart disease as a specific diagnosis, and for cerebrovascular disease as a specific diagnosis in particular, in both cohorts (Table 8.6).[41] The potential for confounding by passive smoke exposure in this study is outlined above, but the more recent INTERHEART case-control study, involving 12,133 cases and 14,435 controls across 52 countries, provides further evidence.[62] This study controlled for a number of potential confounding variables, including exposure to second-hand smoke, and estimated an adjusted odds ratio of 1.57 (95% CI 1.24 to 2.00) for tobacco chewers, adjusted for smoking status. In analysis of exclusive tobacco chewers (excluding smokers), the odds ratio for myocardial infarction was significantly increased at 2.23 (95% CI 1.41 to 3.52) compared with never tobacco users. Adjustment for other potential confounding variables such as diabetes, obesity, hypertension, exercise, and diet had little effect. The risks of acute myocardial infarction associated with smokeless and other types of tobacco use from the INTERHEART study are shown in Fig 8.5.[62] There were insufficient numbers of users included in the study to draw conclusions about the product-specific effects of snuff or paan use.

A prospective incident case-control study of the cardiovascular risks from snus in Sweden compared 525 male cases of myocardial infarction and 1,798 matched controls.[63] This study found no increased risk of myocardial infarction or sudden cardiac death among exclusive snuff users, whereas smokers' risks were two to three times those of never smokers.

The authors of the INTERHEART study hypothesised that the mechanisms whereby oral tobacco may increase the risk of myocardial infarction could involve

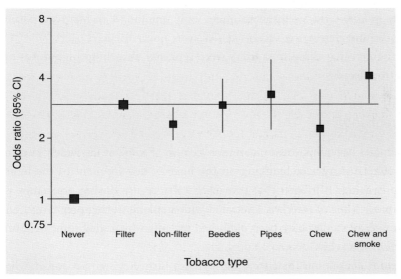

Fig 8.5 Risk of acute myocardial infarction associated with type of tobacco used. Odds ratio for current smokers = 2·95 (95% CI 2·77–3·14) indicated by broken horizontal line. Never = never smokers; Filter = filter cigarettes; Non-filter = non-filter cigarettes; Beedies = smoking beedies alone; Pipes = smoking pipes/cigars; Chew = chewing tobacco alone; Chew and smoke = both chewing and smoking tobacco. Reproduced from Teo *et al* with permission from Elsevier, copyright 2006.[62]

arterial damage and short-term increases in blood pressure caused by nicotine.[64,65] If the mechanism is mediated largely via nicotine effects, then these findings would be likely to apply to all forms of smokeless tobacco, as well as to long-term use of nicotine replacement therapy. The main weakness of the INTERHEART study's analysis of smokeless tobacco use was that many of those included in the analyses were likely to also chew areca nut and other additives along with their tobacco. Given the known cardiovascular effects of areca nut, it is unclear how much of the excess risk in the INTERHEART study can be attributed to tobacco.[13]

In conclusion, the evidence remains mixed but three large studies have found increased risks of cardiovascular disease in smokeless tobacco users, as compared with never tobacco users.[41,51,62] Where these studies reported the cardiovascular risks associated with smoking, these were always greater than those associated with smokeless tobacco use. The possibility remains that a part of these effects of smokeless tobacco use is due to confounding variables (such as passive smoke exposure) or to concomitant use of areca nut. However, the most plausible interpretation is that long-term smokeless tobacco use is associated with a modest increased risk of serious cardiovascular events (and specifically myocardial infarction). It is also plausible that this elevated risk is due to the effects of nicotine,[64,65] though the potential contributions of other components of smokeless tobacco have not, to date, been investigated. However, of six studies of risks of myocardial infarction in long-term Swedish snus users, only one found an increased risk,[51] and five did not detect any increased risk over never tobacco users,[57,58,60–63] despite the fact that these snus users were typically absorbing high doses of nicotine over decades.

8.4.6 Pregnancy

Animal studies have implicated nicotine as a cause of some of the widely known adverse effects of tobacco smoking on the healthy development of the fetus during pregnancy. It follows that smokeless tobacco use during pregnancy is likely to incur some of the risks associated with smoking during pregnancy.[66] A small number of studies have been published recently which shed some light on the effects of smokeless tobacco on pregnancy.

England *et al* examined data from the Swedish Birth Register for women who delivered babies during 1999–2000.[67] The study compared 789 snus users with 11,240 cigarette smokers and 11,495 women not using any tobacco. Compared with non-users, adjusted mean birth weight was reduced in snuff users by 39 g (95% CI 6 g to 72 g) and in smokers by 190 g (95% CI 178 g to 202 g). Preterm delivery was increased in snuff users (adjusted odds ratio 1.98; 95% CI 1.46 to

2.68) and smokers (adjusted odds ratio 1.57; 95% CI 1.38 to 1.80). Pre-eclampsia was reduced in smokers (adjusted odds ratio 0.63; 95% CI 0.53 to 0.75), but increased in snuff users (adjusted odds ratio 1.58; 95% CI 1.09 to 2.27).

Steyn et al studied the patterns and effects of maternal snuff use, cigarette smoking and exposure to environmental tobacco smoke during pregnancy on birth weight and gestational age in women living in Johannesburg and Soweto in 1990.[68] A cohort of 1,593 women provided information about their own and household members' use of tobacco products during pregnancy. Women who smoked cigarettes or used snuff during pregnancy accounted for 6.1% and 7.5% of the study population respectively. The birth weight reduction adjusted for possible confounders was 137 g (95% CI 26.6 g to 247.3 g) for cigarette smokers and 17.1 g (95% CI -69.5 to 102.7 g) for snuff users, when compared with the birth weight of non-tobacco users. They concluded that snuff use was not associated with low birth weight.

Gupta and Subramoney conducted a population-based prospective cohort study to investigate whether smokeless tobacco use was associated with increased risks for stillbirth in India.[69] The adjusted risk for stillbirth in users was 2.6 (95% CI 1.4 to 4.8). Most women used mishri (a pyrolysed tobacco product often used as dentifrice), and there was a dose-response relationship between the daily frequency of use and stillbirth risk.

Overall, these epidemiological studies are consistent with the evidence from animal studies that exposure to nicotine via smokeless tobacco in pregnancy is harmful to the fetus, causing a reduction in birth weight of about 20% of that caused by smoking, and increasing risks of pre-eclampsia and stillbirth.

8.4.7 Periodontal disease

There is clear evidence that smokeless tobacco use causes smokeless tobacco keratosis, gum erosion and increases the risk of periodontal disease, although typically to a lesser extent than cigarette smoking.[16,70–72]

8.5 Health effects of smokeless tobacco compared with cigarettes

Very few studies have directly compared the health risks of smokeless tobacco use with those of cigarette smoking. Roth et al reviewed studies comparing the risks of smoking with those of Swedish snus.[73] They found only seven studies in which direct comparisons were made. Although few in number, these studies do provide quantitative evidence that, for certain health outcomes, the health risks associated with snus are lower than those associated with smoking. Specifically, this is true for lung cancer (based on one study), for oral cancer (based on one study), for gastric cancer (based on one study), for cardiovascular disease (based on three of four studies), and for all-cause mortality (based on one study).

Another way of indirectly assessing the magnitude of risks is to compare the published hazard ratios for specific diseases in smokeless users and smokers in the CPS-II study in the United States. Here we find that smokers have relative risks of death from oral cancers of 10.9 to 14.6, relative risks of lung cancer death of 23.3, and relative risk of death from COPD of 10.6,[74] as compared with adjusted hazard ratios of 0.9, 2.0, and 1.3 respectively for the most directly comparable diagnostic categories in smokeless tobacco users (all of these relative to never tobacco users).[41]

Quantification of the magnitude of the difference in risks could be aided by direct comparison using the same diagnostic categories and risk measure, but it is very clear that, for most of the major health effects of tobacco, smoking is many times more dangerous than smokeless tobacco use. Perhaps the main exceptions are pre-eclampsia, because other components in tobacco smoke appear to have a mild preventive effect,[67] and cardiovascular disease, in which the risks from smoking, although certainly greater than those from smokeless use, may be in a similar range of risk. As discussed in section 8.4.5, cardiovascular risk estimates vary widely, but are typically slightly greater than those of never tobacco users and consistently less than those of smokers. For example, in the CPS studies in the United States, relative risks for smokers compared with never smokers ranged from 1.5 for ischaemic heart disease in men over 65 to 7.1 for aortic aneurysm in women.[74] As shown in Table 8.6, the adjusted hazard ratios for male smokeless users compared with never tobacco users ranged from 1.12 for coronary heart disease to 1.46 for cerebrovascular disease.

In an exercise carried out to estimate the relative hazards of snus use and smoking, a panel of nine experts was asked to determine expert opinions of mortality risks associated with use of low-nitrosamine smokeless tobacco (LN-SLT) marketed for oral use.[11] They concluded that, in comparison with smoking, there is at least a 90% reduction in the relative risk of LN-SLT use. In reporting the results of a subsequent study using a similar design, the authors concluded that the introduction of a new LN-SLT product in the United States under strict regulations would increase SLT use, but reduce overall smoking prevalence. They said, 'This reduction would likely yield substantial health benefits but uncertainties surround the role of marketing and other tobacco control policies.'[75] Much of the evidence for this conclusion is based on the experience in Sweden of snus becoming more popular than smoking in Swedish men, and being used by a significant proportion of Swedish smokers in the process of quitting smoking.[4,76,77]

One study found that among men who made attempts to quit smoking, snus was the most commonly used cessation aid, being used by 24% on their latest quit attempt (meaning it is used more frequently as a cessation aid than the total use of all the nicotine replacement products).[77] Importantly, the data from Sweden also suggest that rather than snus acting as a gateway to smoking for young people, those people who take up snus use are actually significantly less likely to become regular smokers.[77-79] In the northern part of Sweden, where snus use is particularly popular, the effects on reducing smoking have been particularly marked. One recent study reported that the prevalence of daily smoking among men in northern Sweden fell from 19% in 1986 to only 9% (CI 7.0% to 11%) in 1999, and to only 3% (CI 0.1% to 5.4%) in men age 25–34 years; while the prevalence of exclusive snus use increased to 27% (95% CI 24 to 30%) and 34% (95% CI 27 to 42%) respectively.[80]

8.5.1 Health effects in ex-smokers who switch to smokeless tobacco

One recent study using the CPS-II cohort compared tobacco-related disease mortality among 4,443 men who switched from cigarettes to smokeless tobacco ('switchers'), with those of 111,952 men who quit smoking completely without using any other tobacco ('quitters').[81] After 20 years of follow-up, switchers had a higher rate of death from any cause (hazard ratio 1.08; 95% CI 1.01 to 1.15), lung cancer (hazard ratio 1.46; CI 1.24 to 1.73), coronary heart disease (hazard ratio 1.15; 1.00 to 1.29), cancer of the oral cavity and pharynx (hazard ratio 2.5; CI 1.2 to 5.7 based on only seven deaths in switchers) and stroke (hazard ratio 1.24; 1.01 to 1.53), compared with those who quit tobacco entirely. Interestingly, this study also found almost significantly increased risk of death from COPD in switchers (hazard ratio 1.32; CI 0.96 to 1.78). The analyses were also separated into chew and snuff users. For snuff users (27% of switchers), the adjusted hazard ratios for all cause mortality (1.11), coronary heart disease (1.12) and stroke (0.89) were not significant elevated, and were lower than those for lung cancer (hazard ratio 1.75; CI 1.2 to 2.5) and COPD (hazard ratio 1.68; CI 0.9 to 3.3).

The switchers and quitters differed significantly in some important baseline characteristics (for example, 30% of the switchers had less than a high school education compared with 13% of quitters), but the results summarised above were adjusted for a range of relevant demographic and behavioural factors, including previous smoking rate, years of smoking, fruit intake, body mass index and other factors recorded at recruitment in 1982. The authors acknowledged the possibility that residual confounding (for example, exposure to second-hand smoke or subsequent relapse to smoking) could account for some or all of the increased risk in switchers. However, they concluded that this study, together with the other CPS-II data, suggests that using smokeless tobacco compares unfavourably with both complete tobacco cessation and never use of tobacco products. Unfortunately, neither this nor the other papers from the CPS studies have reported a comparison of relative risks in smokeless users and smokers. This would be informative for smokers who feel they either cannot, or do not want to, quit tobacco, but would like to significantly reduce their risks from tobacco use.

8.5.2 Smokeless tobacco: a gateway to smoking or a means of quitting smoking?

Given that smokeless tobacco is much less harmful than smoking, the issue of how smokeless tobacco may influence smoking rates is relevant to judging its overall health impact. If people who start using smokeless tobacco are at a higher risk of starting smoking or a lower risk of quitting smoking (all other things being equal), this would add significantly to the health impact of smokeless tobacco use. Conversely, if smokeless tobacco use reduces risk of smoking or helps smokers to quit, that should also be considered in weighing its harmfulness, both for the individual and for public health.

Unfortunately, there is very little available information on patterns of use of smokeless tobacco and effects on smoking prevalence outside Sweden. Rodu and Godshall have reviewed the evidence from national surveys in the United States.[21] They reported that the 1986 US Adult Use of Tobacco Survey found that 7% (1.7 million) of male ex-smokers had used smokeless tobacco to help them quit smoking cigarettes, as compared with 1.7% of male ex-smokers (405,000) who had used organised smoking cessation programmes to help them quit smoking. The 1991 US National Health Interview (NHI) Survey revealed that a third of adult smokeless tobacco users were former cigarette smokers (around 1.8 million). The 1998 NHI Survey revealed that 5.8% of daily snuff users reported quitting smoking cigarettes within the past year; that daily snuffers were three times more likely to report being former smokers than were never snuff users; and that daily snuff users were four times more likely to have quit smoking in the past year than never snuff users.

The question of whether smokeless tobacco use acts as a gateway to smoking in the United States has been the subject of heated debate. Kozlowski *et al* reported that a maximum of 23% of US young smokeless tobacco users progressed into smoking, based on the finding that 35% of 23–34-year-old smokeless users had never smoked, and 42% smoked before they used smokeless tobacco.[82] They also found that those who used cigarettes before moist snuff were 2.1 times more likely to have quit smoking (95% CI 1.21 to 6.39) than cigarette-only users. They concluded that the large majority of US smokeless users do not in fact progress to smoking. However, a study of recruits to the US Army by Haddock *et al* and a longitudinal study of almost 4,000 US male adolescents by Tomar both found that young men who started using smokeless tobacco were approximately twice and 3.5 times respectively more likely to smoke subsequently.[83,84] While these findings must be interpreted cautiously since use of smokeless tobacco and progression to smoking are likely to be biased by a range of cultural and other factors, there is evidently some uncertainty over the extent of gateway use in these and other US studies.

Recent studies in Sweden, however, appear to be consistent in finding that Swedes who start using snus are less likely to become smokers and Swedish smokers who start using snus are more likely to quit smoking.[77–80] The study by Ramström *et al* found that the odds of initiating daily smoking were significantly lower for men who had started using snus than for those who had not (odds ratio 0.28; 95% CI 0.22 to 0.36) (Fig 8.6).[77] Among male primary smokers, 28% started secondary daily snus use and 73% did not. Of those secondary snus users, 88% had ceased daily smoking completely by the time of the survey compared with 56% of those primary daily smokers who never became daily snus users, (odds ratio 5.7; 95% CI 4.9 to 8.1). Among men who made attempts to quit smoking, snus was the most commonly used cessation aid, being used by 24% on their latest quit attempt.

One potential reason why more people have switched from smoking to smokeless tobacco in Sweden than the United States is that in Sweden the public health authorities have updated their information on health risks according to the latest scientific evidence,[85] whereas in the United States the public have been informed that smokeless tobacco is just as harmful as smoking cigarettes.[86,87] In an environment in which smokers are informed and believe that smokeless tobacco is just as harmful to health as smoking, one would not expect many to switch for health reasons.

A recent comprehensive systematic review and meta-analysis of the health effects of snus concluded that:

The evidence suggests that the harm of using snus relative to non-tobacco is significantly less than found for smoking with regard to cancer of the head, neck and gastrointestinal region, and cardiovascular disease events.[88]

Snus is currently banned in Australia. Gartner *et al* used the best available data and expert reviews to estimate the likely health impact of snus in Australia.[89] They conclude that snus is likely to produce a net benefit to population health, with the size of the benefit dependent on how many inveterate smokers switch to snus.

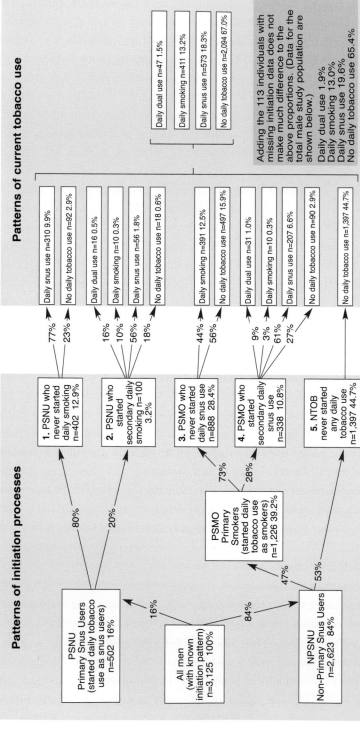

Fig 8.6 Pathways of male tobacco use in Sweden (percentages in boxes based on all men ages 16–79). The left section shows the proportions of men who started daily snus use, smoking, both (and if so, in what order) or no tobacco use. The central section shows, for each initiation category, the proportions of men who continued or quit each type of tobacco use that was initiated. The right section shows the proportions of men following 13 different pathways to current tobacco use. NPSNU = non-primary snus users; NTOB = never tobacco users; PSNU = primary snus users; PSMO = primary smokers. From Ramstrom and Foulds;[77] reproduced with permission from the BMJ Publishing Group Ltd.

The epidemiology of tobacco use in Sweden suggests that if the public is offered a substantially less harmful smokeless tobacco product along with access to accurate information on relative risks, a substantial proportion can switch to the less harmful product. This has clear implications for public health.[90]

8.6 Conclusions

▸ Smokeless tobacco is not a single product, but rather a summary term for a range of different tobacco products which deliver nicotine without combustion.

▸ Smokeless tobacco products differ substantially in their risk profile in approximate relation to the content of toxins in the tobacco.

▸ In some parts of the world (particularly Asia), smokeless tobacco is commonly mixed with other products that are themselves harmful.

▸ On toxicological and epidemiological grounds, some of the Swedish smokeless products appear to be associated with the lowest potential for harm to health.

▸ These Swedish smokeless products appear to increase the risk of pancreatic cancer, and possibly of cardiovascular disease, particularly myocardial infarction.

▸ Some smokeless tobacco products also increase the risk of oral cancer, but if true of Swedish smokeless tobacco, the magnitude of this effect is small.

▸ All of the above hazards are of a lower magnitude than those associated with cigarette smoking.

▸ Smokeless tobacco products have little or no effect on the risk of chronic obstructive pulmonary disease or lung cancer.

▸ Therefore, in relation to cigarette smoking, the hazard profile of the lower risk smokeless products is very favourable.

▸ Smokeless tobacco use by pregnant women is harmful to the unborn fetus, but the hazard of smokeless use relative to maternal cigarette smoking is not clearly established.

▸ In Sweden, the available low-harm smokeless products have been shown to be an acceptable substitute for cigarettes to many smokers, while 'gateway' progression from smokeless to smoking is relatively uncommon.

▸ Smokeless tobacco, therefore, has potential application as a lower hazard alternative to cigarette smoking.

▸ The applicability of smokeless tobacco as a substitute for cigarette smoking if made available to populations with no tradition of smokeless use is not known.

References

1   Collishaw NE, Lopez AD. The tobacco epidemic: a global public health emergency. Geneva: World Health Organization, 1996.

2   Giovino GA. Epidemiology of tobacco use in the United States. *Oncogene*. 2002;21:7326–40

3   US Department of Agriculture. *Tobacco situation and outlook report*. TBS-253. Washington DC: USDA, Commodity Economics Division, Economic Research Service, 2002.

4   Foulds J, Ramstrom L, Burke M, Fagerstrom K. The effect of smokeless tobacco (snus) on public health in Sweden. *Tob Control* 2003;12:349–59.

5   Reddy KS, Perry CL, Stigler MH, Arora M. Differences in tobacco use among young people in urban India by sex, socioeconomic status, age, and school grade: assessment of baseline survey data. *Lancet* 2006;367:589–94.

6   McNeill A, Bedi R, Islam S, Alkhaftib MN, West R. Levels of toxins in oral tobacco products in the UK. *Tob Control* 2006;15:64–7.

7   Idris AM, Ibrahim SO, Vasstrand EN *et al*. The Swedish snus and the Sudanese toombak: are they different? *Oral Oncology* 1998;34:558–66.

8   Roosaar A, Johansson AL, Sandborgh-Englund G, Nyren O, Axell T. A long-term follow-up study on the natural course of snus-induced lesions among Swedish snus users. *Int J Cancer* 2006;119:392–7.

9   Rosenquist K, Wennerberg J, Schildt EB *et al*. Use of moist snuff, smoking and alcohol consumption in the aetiology of oral and oropharyngeal cell carcinoma. A population-based case-control study in southern Sweden. *Acta Oto-Laryngologica* 2005;125:991–8.

10 Zatterstrom UK, Svennson M, Sand L, Nordgren H, Hirsch JM. Oral cancer after using Swedish snus (smokeless tobacco) for 70 years – a case report. *Oral Dis*;2004 10:50–3.

11 Levy DT, Mumford EA, Cummings KM *et al*. The relative risks of a low-nitrosamine smokeless tobacco product compared with smoking cigarettes: estimates of a panel of experts. *Cancer Epidemiol Biomarkers Prev* 2004;13:2035–42.

12 US National Cancer Institute and Centers for Disease Control. *Smokeless tobacco*. Factsheet. Third International Conference on Smokeless Tobacco, 2002.

13 Chu NS. Neurological aspects of areca and betel chewing. *Addict Biol* 2002; 7:111–4.

14 Subramanian SV, Nandy S, Kelly M, Gordon D, Davey Smith G. Patterns and distribution of tobacco consumption in India: cross sectional multilevel evidence from the 1998-9 national family health survey. *BMJ* 2004;3;328:801–6.

15 Rani M, Bonu S, Jha P, Nguyen SN, Jamjoum L. Tobacco use in India: prevalence and predictors of smoking and chewing in a national cross sectional household survey. *Tob Control* 2003;12:e4.

16 Winn DM. Tobacco use and oral disease. *J Dent Educ* 2001;65:306–12.

17 Winn DM, Blot WJ, Shy CM *et al*. Snuff dipping and oral cancer among women in the southern United States. *N Engl J Med* 1981;304:745–9.

18 Sapundzhiev N, Werner JA. Nasal snuff: historical review and health related aspects. *J Laryngol Otol* 2003;117:686–91.

19 Sreedharan S, Kamath MP, Khadilkar U *et al*. Effect of snuff on nasal mucosa. *Am J Otolaryngol* 2005;26:151–6.

20 Nelson DE, Mowery P, Tomar S *et al*. Trends in smokeless tobacco use among adults and adolescents in the United States. *Am J Public Health* 2006;96:897–905.

21 Rodu B, Godshall WT. Tobacco harm reduction: an alternative cessation strategy for inveterate smokers. *Harm Reduct J* 2006;3:37.

22 US Department of Agriculture Economic Research Service: *Briefing rooms: tobacco*, May 2007. www.ers.usda.gov/Briefing/Tobacco/ (accessed 7 August 2007).

23 Foulds J, Russell MA, Jarvis MJ, Feyerabend C. Nicotine absorption and dependence in unlicensed lozenges available over the counter. *Addiction* 1998;93:1427–31.

24 WHO International Agency for Research on Cancer. *Volume 89: Smokeless tobacco and tobacco-specific nitrosamines.* IARC monograph on the evaluation of carcinogenic risk of chemicals to humans. Lyon: IARC, 2006.

25 Brunnemann KD, Hoffmann D. Chemical composition of smokeless tobacco products. In: National Cancer Institute, *Smokeless tobacco or health: an international perspective.* Smoking and Tobacco Control Monograph No. 2. Bethesda, MD: US Department of Health and Human Services, National Institutes of Health, 1993:96–106. cancercontrol.cancer.gov/tcrb/monographs/2/index.html (accessed 7 August 2007).

26 Brunnemann KD, Qi J, Hoffmann D. *Aging of oral moist snuff and the yields of tobacco specific n-nitrosamines (TSNA).* Progress report prepared for the Massachusetts Tobacco Control Program. Boston MA: Department of Public Health, June 2001. www.tobacco.org/News/010622BostonRe.html (accessed 7 August 2007).

27 Idris AM, Nair J, Oshima H *et al.* Unusually high levels of carcinogenic nitrosamines in Sudan snuff (toombak). *Carcinogenesis* 1991;12:1115–8.

28 Hoffmann D, Djordjevic MV, Fan J *et al.* Five leading U.S. commercial brands of moist snuff in 1994: assessment of carcinogenic N-nitrosamines. *J Natl Cancer Inst* 1995;87: 1862–9.

29 Osterdahl BG, Jansson C, Paccou A. Decreased levels of tobacco-specific N-nitrosamines in moist snuff on the Swedish market. *J Agric Food Chem* 2004;52:5085–8.

30 Rodu B, Jansson C. Smokeless tobacco and oral cancer: a review of the risks and determinants. *Crit Rev Oral Biol Med* 2004;15:252–263.

31 Stepanov I, Hecht SS, Ramakrishnan S, Gupta PC. Tobacco-specific nitrosamines smokeless tobacco products marketed in India. *Int J Cancer* 2005;116:16–19.

32 Stepanov I, Jensen J, Hatsukami D, Hecht SS. Tobacco-specific nitrosamines in new tobacco products. *Nicotine Tob Res* 2006;8:309–13.

33 Merchant A, Husain SS, Hosain M *et al.* Paan without tobacco: an independent risk factor for oral cancer. *Int J Cancer* 2000;86:128–31.

34 Balaram P, Sridhar H, Rajkumar T *et al.* Oral cancer in southern India: the influence of smoking, drinking, paan-chewing and oral hygiene. *Int J Cancer* 2002;98:440–5.

35 Lee KW, Kuo WR, Tsai SM *et al.* Different impact from betel quid, alcohol and cigarette: risk factors for pharyngeal and laryngeal cancer. *Int J Cancer* 2005;117:831–6.

36 WHO International Agency for Research on Cancer. *Volume 37: Tobacco habits other than smoking: betel quid and areca nut chewing and some related nitrosamines.* IARC monograph on the evaluation of carcinogenic risk of chemicals to humans. Lyon: IARC, 1985.

37 Thongsuksai P, Boonyaphiphat P, Sriplung H, Sudhikaran W. p53 mutations in betel-associated oral cancer from Thailand. *Cancer Lett* 2003;201:1–7.

38 Critchley JA, Unal B. Health effects associated with smokeless tobacco: a systematic review. *Thorax.* 2003;58:435–43.

39 Critchley JA, Unal B. Is smokeless tobacco a risk factor for coronary heart disease? A systematic review of epidemiological studies. *Eur J Cardiovasc Prev Rehabil* 2004;11:101–12.

40 Stratton K, Shetty P, Wallace R, Bondurant S (eds), *Clearing the smoke: assessing the science base for tobacco harm reduction.* Institute of Medicine, National Academy of Sciences. Washington DC: National Academy Press, 2001.

41   Henley SJ, Thun MJ, Connell C, Calle EE. Two large prospective studies of mortality among men who use snuff or chewing tobacco (United States). *Cancer Causes Control* 2005;16:347–58.

42   Neville BW, Day TA. Oral cancer and precancerous lesions. *CA Cancer J Clin* 2002;52;195–215.

43   Accortt NA, Waterbor JW, Beall C, Howard G. Cancer incidence among a cohort of smokeless tobacco users (United States). *Cancer Causes Control* 2005;16:1107–15.

44   Luo J, Ye W, Zendehdel K *et al.* Oral use of Swedish moist snuff and risk for cancer of the mouth, lung and pancreas among male construction workers. *Lancet* 2007;369:2015–20.

45   Prokopczyk B, Hoffmann D, Bologna M *et al.* Identification of tobacco-derived compounds in human pancreatic juice. *Chem Res Toxicol* 2002;15:677–85.

46   Stepanov I, Hecht SS. Tobacco-specific nitrosamines and their pyridine-N-glucuronides in the urine of smokers and smokeless tobacco users. *Cancer Epidemiol Biomarkers Prev* 2005;14:885–91.

47   Bofetta P, Aagnes B, Weiderpas E, Andersen A. Smokeless tobacco use and risk of cancer to the pancreas and other organs. *Int J Cancer* 2005;114:992–5.

48   Alguacik J, Silverman DT. Smokeless and other noncigarette tobacco use and pancreatic cancer: a case-control study based on direct interviews. *Cancer Epidemiol Biomarkers Prev* 2004;13:55–8.

49   Hecht SS. Reply to Foulds and Ramstrom. *Cancer Causes Control* 2006;17:859–60.

50   Foulds J, Ramstrom R. Causal effects of smokeless tobacco on mortality in CPS-I and CPS-II? *Cancer Causes Control* 2006;17:227–8.

51   Bolinder G, Alfredsson L, Englund A, de Faire U. Smokeless tobacco use and increased cardiovascular mortality among Swedish construction workers. *Am J Public Health* 1994;84:399–404.

52   Gajalakshmi V, Hung RJ, Mathew A *et al.* Tobacco smoking and chewing, alcohol drinking and lung cancer risk among men in southern India. *Int J Cancer* 2003;107:441–7.

53   Ye W, Ekstrom AM, Hansson LE, Bergstrom R, Nyren O. Tobacco, alcohol and the risk of gastric cancer by sub-site and histologic type. *Int J Cancer* 1999;83:223–9.

54   Lewin F, Norell SE, Johansson H, Gustavsson P, Wennerberg J. Smoking tobacco, oral snuff, and alcohol in the etiology of squamous cell carcinoma of the head and neck. *Cancer* 1998; 82:1367–74.

55   Schildt E-B, Eriksson M, Hardell L, Magnusson A. Oral snuff, smoking habits and alcohol consumption in relation to oral cancer in a Swedish case-control study. *J Cancer* 1998;77:341–6

56   Lagergren J, Bergstrom, Lindgren A, Nyren O. The role of tobacco, snuff and alcohol use in the aetiology of cancer of the oesophagus and gastric cardia. *J Cancer* 2000;85:340–6.

57   Huhtasaari F, Asplund K, Lundberg V, Stegmayr B, Wester PO. Tobacco and myocardial infarction: is snuff less dangerous than cigarettes? *BMJ* 1992;305:1252–6.

58   Huhtasaari F, Lundberg V, Eliasson M, Janlert U, Asplund K. Smokeless tobacco as a possible risk factor for myocardial infarction: a population-based study in middle-aged men. *J Am Coll Cardiol* 1999;34:1784–90.

59   Asplund K. Smokeless tobacco and cardiovascular disease. *Prog Cardiovasc Dis* 2003;45: 383–94.

60   Hergens MP, Ahlbom A, Andersson T, Pershagen G. Swedish moist snuff and myocardial infarction among men. *Epidemiology* 2005;16:12–16.

61 Johansson SE, Sundquist K, Qvist J, Sundquist J. Smokeless tobacco and coronary heart disease: a 12-year follow-up study. *Eur J Cardiovasc Prev Rehabil* 2005;12:387–92.

62 Teo KK, Ounpuu S, Hawken S *et al*; INTERHEART Study Investigators. Tobacco use and risk of myocardial infarction in 52 countries in the INTERHEART study: a case-control study. *Lancet* 2006;368:647–58.

63 Wennberg P, Eliasson M, Hallmans G *et al*. The risk of myocardial infarction and sudden cardiac death amongst snuff users with or without a previous history of smoking. *J Int Med;* 13 Apr 2007, doi: 10.1111/j.1365-2796.2007.01813.x

64 Bolinder G, Norén A, de Faire U, Wahren J. Smokeless tobacco use and atherosclerosis: an ultrasonographic investigation of carotid intima media thickness in healthy middle-aged men. *Atherosclerosis* 1997;132:95–103.

65 Bolinder G, de Faire U. Ambulatory 24-h blood pressure monitoring in healthy, middle-aged smokeless tobacco users, smokers, and non tobacco users. *Am J Hypertens* 1998;11:1153–63.

66 Ernst M, Moolchan ET, Robinson ML. Behavioral and neural consequences of prenatal exposure to nicotine. *J Am Acad Child Adolesc Psychiatry* 2001;40:630–41.

67 England LJ, Levine RJ, Mills JL *et al*. Adverse pregnancy outcomes in snuff users. *Am J Obstet Gynecal* 2003;189:939–43.

68 Steyn K, de Wet T, Nel H, Yach D. The influence of maternal cigarette smoking, snuff use and passive smoking on pregnancy outcomes: the Birth To Ten Study. *Paediatr Perinat Epidemiol* 2006;20:90–9.

69 Gupta PC, Subramoney S. Smokeless tobacco use and risk of stillbirth: a cohort study in Mumbai, India. *Epidemiology* 2006;17:47–51.

70 Wickholm S, Soder PO, Galanti MR, Soder B, Klinge B. Periodontal disease in a group of Swedish adult snuff and cigarette users. *Acta Odontol Scand* 2004;62:333–8.

71 Fisher MA, Taylor GW, Tilashalski KR. Smokeless tobacco and severe active periodontal disease, National Health and Nutrition Examination Survey (NHANES) III. *J Dent Res* 2005;84:705–10.

72 Tomar SL, Asma S. Smoking-attributable periodontitis in the United States: findings from National Health and Nutrition Examination Survey (NHANES) III. *J Periodontol* 2000;71:743–51.

73 Roth D, Roth AB, Liu X. Health risks of smoking compared with Swedish snus. *Inhal Toxicol* 2005;17:741–8.

74 Thun MJ, Myers DG, Day-Lally C *et al*. Age and exposure-response relationships between cigarette smoking and premature death in Cancer Prevention Study II. In: Shopland DR, Burns DM, Garfinkel L, Samet JM (eds), *Changes in cigarette-related disease risks and their implications for prevention and control*. Smoking and Tobacco Control Monograph No. 8. NIH Publication number 97-4213. Bethesda, MD: National Cancer Institute, 1997.

75 Levy DT, Mumford EA, Cummings KM *et al*. The potential impact of a low-nitrosamine smokeless tobacco product on cigarette smoking in the United States: estimates of a panel of experts. *Addict Behav* 2006;31:1190–200.

76 Gilljham H, Galanti MR. Role of snus (oral moist snuff) in smoking cessation and smoking reduction in Sweden. *Addiction* 2003;98:1183–9.

77 Ramström LM, Foulds J. The role of snus (smokeless tobacco) in initiation and cessation of tobacco smoking in Sweden. *Tob Control* 2006;15:210–4.

78 Furberg H, Bulik C, Lerman C *et al*. Is Swedish snus associated with smoking initiation or smoking cessation? *Tob Control* 2005;14:422–4.

79   Rodu B, Nasic S, Cole P. Tobacco use among Swedish schoolchildren. *Tob Control* 2005;14:405–8.

80   Stegmayr B, Eliasson M, Rodu B. The decline of smoking in northern Sweden. *Scand J Public Health* 2005;33:321–4.

81   Henley SJ, Connell CJ, Richter P *et al.* Tobacco-related disease mortality among men who switched from cigarette to spit tobacco. *Tob Control* 2007;16:22–8

82   Kozlowski LT, O'Connor RJ, Edwards BQ, Flaherty BP. Most smokeless tobacco use is not a casual gateway to cigarettes: using order of product use to evaluate causation in a national US sample. *Addiction* 2003;98:1077–85.

83   Haddock CK, Weg MV, Debon M, Klesges R *et al.* Evidence that smokeless tobacco use is a gateway for smoking initiation in young adult males. *Prev Med* 2001;32:262–7.

84   Tomar S. Is use of smokeless tobacco a risk factor for cigarette smoking? The U.S. experience. *Nicotine Tob Res* 2003;5:561–9.

85   Ahlbom A, Olsson UA, Pershagen G. *Health hazards of moist snuff.* Sweden: National Board of Health and Welfare, 1997;11:1–30.

86   Carmona R. *Can tobacco cure smoking? A review of harm reduction.* Testimony before the US House Subcommittee on Commerce, Trade, and Consumer Protection. June 3, 2003.

87   Kozlowski LT, Edwards BQ. 'Not safe' is not enough: smokers have a right to know more than there is no safe tobacco product. *Tob Control* 2005;14(Suppl 2):3–7.

88   Broadstock M. *Systematic review of the health effects of modified smokeless tobacco products.* New Zealand Health Technology Assessment Report 2007;10:1–110. nzhta.chmeds.ac.nz/publications.htm (accessed 7 August 2007).

89   Gartner CE, Hall WD, Vos T, Bertram MY, Wallace AL, Lim SS. Assessment of Swedish snus for tobacco harm reduction. *Lancet* 2007;369:2010–4.

90   Foulds J, Kozlowski L. Snus – what should the public health response be? *Lancet* 2007;369:1976–8.

9 | Current nicotine product regulation

9.1 Introduction

The range of nicotine and tobacco products available in most countries is extensive, and includes pure medicinal nicotine products; chewing tobacco, snuff and other smokeless tobacco products; raw tobacco for smoking in hand-rolled cigarettes or pipes; and cigarettes and cigars (see Chapter 5). Currently, there is no overarching regulatory framework for these products in the UK or any other country. In addition, there is no systematic regulatory process applied across the production of tobacco products from manufacturers to distributors, wholesalers, retailers and marketers, although some recent attempts have been made to control the supply chain through government initiatives to control smuggling. In recent years, the UK and other countries have implemented a broad range of tobacco control strategies,[1,2] but regulation of the product itself has received relatively little attention or resource. This may be partly due to a lack of clarity as to whether it is possible to make cigarettes less harmful and, if so, how best to do this. It is also unclear what role other tobacco and nicotine products can play in reducing the health burden caused by tobacco use in the UK.

Cigarettes were developed and marketed in the UK and elsewhere decades before the dangers of smoking were known, and were therefore widely used long before a regulatory response proportionate to their risk could be developed. As a result, cigarettes have remained largely exempt from consumer protection legislation, and tobacco product regulations were introduced gradually in a piecemeal way over the years by various regulatory bodies or government departments with responsibility for different aspects and categories of the product. For example, at present some tobacco product regulations, such as machine-measured tar, nicotine and carbon monoxide yields of cigarettes and the supply of oral tobacco, are decided at EU level by European Directives and apply across the UK and other EU countries. Some, such as taxation, smuggling and advertising controls other than those with cross border effects, are decided at UK level. Others, such as smoke-free legislation, are regulated by the devolved administrations in Scotland, Wales and Northern Ireland. Medicinal nicotine products are regulated by the Medicines and Healthcare products Regulatory Agency (MHRA), which has a UK-wide remit.

For the purposes of this chapter we have divided the range of nicotine and tobacco products into four categories: smoked tobacco, smokeless tobacco, new tobacco product formulations and potential reduced exposure products (PREPs), and 'clean' (typically medicinal) nicotine products. Brief details follow on the main types of regulations that apply to these different product categories, grouped largely according to product, price, promotion and place (illustrated in Table 9.1). The chapter then explores the options for improving and optimising the regulatory framework for these products, predominantly from a UK (and EU) perspective.

9.2 Smoked tobacco products

Smoked tobacco products include cigarettes, hand-rolling tobacco, pipes and cigars. Before the emergence of evidence of a link between smoking and lung cancer[3] and

Table 9.1. Range of potential regulations governing tobacco and other nicotine products.

Product	Mode of use	How product is used, for example, smoked or chewed
	Content	Tobacco content, such as nitrosamines and additives
	Design	A range of features such as filters, pressure drop, ventilation, way the tobacco is processed
	Yields	What comes out of the product when used
	Delivery and exposure	The amount of nicotine and other constituents reaching the human body, and the speed of delivery and absorption
Price		Tax structure and price, non-monetary costs, incentives
Promotion		Advertising, sponsorship and marketing
Place		Retail outlet restrictions and accessibility
		Age restrictions
		Packet size restrictions

the Royal College of Physicians (RCP) 1962 report[4] on smoking and health, there was very little regulation of any tobacco product. Aside from tax/duty rules only the sale of tobacco to children was regulated, dating from legislation passed in 1908.

The RCP report set out the need for a comprehensive strategy to reduce smoking, including advertising restrictions, price rises, education about the dangers of smoking, and support for smokers when they attempted to stop smoking. For the most part, however, these were not implemented through legislation. Instead the government entered into an increasingly complex series of 'voluntary agreements' negotiated in private with the tobacco industry. These agreements proved difficult to enforce, left loopholes that were exploited by the industry, and have now been largely discredited as ineffective.[5] From the 1990s, these voluntary agreements were gradually superseded by legislative approaches, often driven by directives from the European Commission.

In 1998, nearly 50 years after the first UK report linking smoking to lung cancer,[3] the UK government published a comprehensive strategy to reduce smoking in England in the form of a White Paper, *Smoking kills*.[1] The stated aim of the strategy was to reduce smoking among children and young people, and help adult smokers – particularly pregnant women and smokers from socioeconomically disadvantaged groups – to quit. The White Paper also recognised that smoking was an addiction. As discussed below, however, despite this progress, the tobacco product itself has remained virtually unregulated. The key regulations applying to smoked tobacco products are as follows.

9.2.1 Price

The retail price of cigarettes reflects the sum of the manufacturers' price, duty and tax, and the retail margin. Increasing the cost of tobacco products through taxation has been widely used as a tobacco control measure in the UK because it both reduces smoking and generates revenue for central government.

The level of duty and taxes on tobacco products (specifically cigarettes, cigars, pipe tobacco, hand-rolling tobacco and chewing tobacco) are set by the Treasury. Initially, excise duty was based only on the tobacco content of cigarettes, but a more complex system was introduced in 1978, when duties on additives and substitutes used in the manufacture of tobacco products were also introduced. Tobacco products in Europe are now subject to both excise tax (levied either as a specific tax or an ad valorem tax) and value added tax (VAT). Specific tax is a fixed sum per cigarette regardless of the value of the product. Ad valorem tax varies in proportion to the value of the goods levied, and thus keeps pace with price inflation. Currently, specific and ad valorem taxes account for approximately equal shares of the excise tax on cigarettes, and total taxation for about 76% of the retail price of cigarettes in the UK. Tax in the UK is now high relative to other European countries (see Fig 9.1),[2] having risen above rates of inflation particularly during the 1990s.

Smuggling and illegal sale of cigarettes has been a significant problem in the UK, and since 2000 a range of policy and regulatory measures have been introduced to reduce the market share[6] of smuggled and counterfeit cigarettes, and hand-rolling tobacco. The Finance Act 2006 included legislation placing obligations on all manufacturers to ensure that cigarettes and hand-rolled tobacco do not evade tobacco duties.[7]

The potential for a taxation strategy to alter consumption has been clearly demonstrated in the UK. In 1978 a supplementary tax was introduced for cigarettes with the highest tar yields (over 20 mg tar) for a period, and resulted in and the near elimination of brands with over 20 mg tar yields from the market.[8] Duties on hand-rolling tobacco were frozen in the UK for a period in the 1990s, because the illicit market share for hand-rolled tobacco had become particularly high. The resulting increase in the price differential between hand-rolling tobacco and manufactured cigarettes only served to exacerbate the problem and there was an

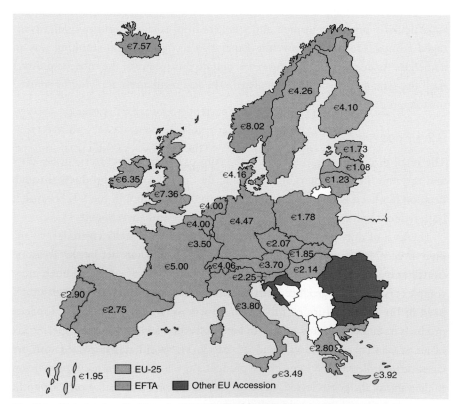

Fig 9.1 Retail price (in Euros) of 20 Marlboro cigarettes in EU countries on 1 January 2006. Adapted from Joosens and Raw, 2006.[2]

increase in the proportion of smokers using hand-rolled tobacco from 10% in 1984 to 25% in 2006.[9] Hand-rolling tobacco in the UK is still taxed at a lower rate compared to cigarettes, only about 66% of the total price being tax.[10]

9.2.2 Promotion

Advertising of tobacco products, defined as 'a product consisting wholly or partly of tobacco and intended to be smoked, sniffed, sucked or chewed', is now banned throughout the UK by law. The Tobacco Advertising and Promotion Act 2002[11] prohibits tobacco advertising on billboards, in print media, by direct mail and through sponsorship. Point of sale advertising is allowed but limited to one A5-sized advertisement, a third of which must be a health warning. The government also has the power under this Act,[11] as yet unused, to introduce further regulations on point of sale displays, such as controlling the number of displays and packets on display, the proximity of tobacco products to children's products like sweets, or potentially by banning all displays of tobacco products at the point of sale.

9.2.3 Product

Regulation of the tobacco product itself is the responsibility of the UK Department of Health, and is currently limited to additives and machine measured yields (see below). A full explanation of how tobacco products are designed, what goes into them and what comes out of them, is not currently required by regulatory authorities in the UK. Articles concerning product regulation are included in the Framework Convention on Tobacco Control,[12] which the UK government has ratified and is thus committed to implement. At present, however, the yields, content and other characteristics of tobacco products in the UK are regulated as follows.

Yields

Since the 1970s, the main regulatory strategy for reducing the harmfulness of cigarettes has been to reduce the machine-smoked tar and nicotine yields they deliver. This process began with a series of voluntary agreements and subsequently through EU regulations, and has been accompanied by introduction of 'low tar' brands with names suggesting reduced harm to the user such as 'light' or 'mild'. The most recent EU directive,[13,14] set maximum yields of tar, nicotine and carbon monoxide for cigarettes.

This approach to regulating cigarettes and its effects on reducing health effects has been extensively reviewed.[15,16] There is now broad recognition that machine-smoked yields do not reflect the exposures sustained in practice by smokers using the product, and that there are minimal, if any, health benefits realised by smoking lower tar and nicotine yield cigarettes. This approach is now recognised as wholly inappropriate as a basis for regulating the harm done by cigarettes,[15] but the regulations remain in force. The European Commission, in their review of the directive in 2005,[17] commented that it did not propose to revise the standards 'until solid evidence shows that better methods exist to replace them'.

There are no regulations concerning the yields of hand-rolling tobacco. An ISO emissions standard has been developed for hand-rolled cigarettes made in a consistent manner and the industry displays this on packs in a number of EU countries such as the Netherlands. The UK has declined to support this because it may be misleading to consumers about the actual real-life yields to smokers. The same argument should be used against displaying yields on packs of manu-factured cigarettes, although this is required by European legislation rather than a voluntary arrangement (see below).

Content

Another approach has been to control the use of additives. Until 1970, few ingredients other than tobacco and paper were used in cigarettes, largely because their use was proscribed. Following the revision of the tax system in 1978, statutory controls on additives became subject to a voluntary agreement with the tobacco industry. The government stated at the time[18] that '... it will amend the Medicines Act 1968 in order to control the use of tobacco substitutes and additives in smoking products in the UK, if the need arises at any time'. This option has not been invoked, and an updated voluntary agreement still applies.

In 2001 a European directive[14] required the disclosure of ingredients for all tobacco products (defined in this legislation as meaning 'products for the purposes of smoking, sniffing, sucking or chewing, inasmuch as they are, even partly, made of tobacco, whether genetically modified or not'). Ingredients added to the products and cigarette papers must now be disclosed by brand name to include the type, quantity and function for each ingredient, and toxicological data. The latter should include effects on health either alone or in combination with other ingredients and addictive effects, for ingredients both in the burnt and unburnt forms. The directive required all member states to ensure that the list of ingredients for each product was made public, though this was qualified by a reference to trade secrecy. How this should be done and by when was not specified. To date, despite the requirements of the European directive to do so, no information on additives in tobacco products has been made public in the UK or in most other European countries. Efforts are currently ongoing with EU member states to create a uniform reporting and publishing standard that would allow consistent public reporting.

Concerns have also been raised as to whether the definition of ingredients and additives is sufficiently comprehensive to pick up all products added to the raw tobacco in the manufacture of cigarettes, and over the manner in which the information is reported.[19] The European Commission has also commented on the incomplete nature of the data it receives on ingredients from member states[17] and is developing harmonised data collection methods based on a common format and improved definitions. Nevertheless, concerns remain about the lack of capacity to assess the data, and further research and analysis has been recommended to create scientifically sound criteria before approval or prohibition of ingredients can be progressed.[19]

There are no regulatory controls on the raw tobacco itself (such as the levels of known carcinogens like nitrosamines) used in combustible products.

Design

There are no regulations concerning the design of tobacco products in the UK. Canada and some US States have regulated the propensity of cigarettes to cause fires by mandating the sale of only fire safe cigarettes. There is currently a campaign for reduced ignition propensity (RIP) legislation to be adopted in the UK, and the Department for Trade and Industry, the Department of Health and the Department for Communities and Local Government have supported setting such a standard.[20]

Product information and health warnings

In response to evidence that machine-smoked tar yields are not valid measures of exposure to the smoker, the European Commission has now required that labels such as 'light' or 'mild' on low tar cigarettes be removed.[14] However, manufacturers are still required to list tar, nicotine and CO yields on cigarette packs,[14] and continue to use other names such as 'smooth', that imply less harm.[21] Health warnings have been required on cigarette packs in the UK since 1971 and were introduced for hand-rolling tobacco (and chewing tobacco products) in 1988. The warning is now required to cover 30% of the front and 40% of the back of cigarette packs.[14] Parties to the FCTC are required to put in place health warnings that are at least 30% but preferably more than 50% of the principal display area of the pack or package within three years of the Treaty coming into force.

In 2003 the European Commission produced rules for the use of graphic warnings on the back of tobacco packs, and subsequently a library of colour photographs for use by member states, but member states could choose whether they required these warnings. To date only Belgium has introduced these, and only on cigarettes. However, the UK government has announced that graphic warnings will be phased in for all tobacco products beginning in the autumn of 2008.

Aside from the above requirements (plus the need for duty paid markings on cigarettes and hand-rolled tobacco and product identification and traceability markings for all tobacco products), the remainder of the tobacco product pack can be used as the industry sees fit. In the wake of advertising bans, the pack remains a key marketing tool for the industry,[22] and can be used to reinforce and maximise positive imagery and associations with a brand and with cigarettes, through careful choice of logos, layout and use of colour.

9.2.4 Place of sale

Regulation of the place of sale is also a responsibility of the Department of Health. No licence is required to sell or distribute tobacco products, which can therefore be distributed and sold from any retail outlet that is VAT registered. In the UK, sale to persons under 16 years of age is prohibited, though there is consistent evidence that childen under 16 years have no difficulty in practice in purchasing cigarettes from shops.[23] In England, Scotland and Wales this minimum age will increase to 18 years from 1 October 2007. It is illegal to sell cigarettes singly or in packets of fewer than 10. An age of sale warning should be displayed wherever tobacco products are sold.

9.3 Smokeless tobacco products

9.3.1 Oral snuff

The supply and sale of oral snuff were first banned by the UK Government under the Consumer Protection Act in 1988 in response to plans by US Tobacco (a leading US manufacturer of smokeless tobacco) to build a factory manufacturing oral snuff products (Skoal Bandits) in Scotland. The law hinged on evidence of the carcinogenicity of the products in the USA, and a concern that US Tobacco was specifically marketing its products at young people.[5] The law was invoked for oral snuff products because, unlike smoked tobacco, they did not yet have an established market. Chewing tobacco was also excluded from the ban because it was in common use among South Asians in the UK.

In July 1992 a European Council Directive[24] prohibited member states from marketing tobacco products for oral use, except those intended to be smoked or chewed. When Sweden subsequently joined the EU it negotiated a special exemption for snus, because there was an established tradition of oral snuff (snus) use in Sweden, going back several decades. In response to this Directive, new regulations to this effect came into force in the UK on 1 January 1993.[25] When the EU Directive was recast in 2001,[14] the ban on oral snuff was retained and the UK regulations cited above remain in force. The ban was challenged in the European Court of Justice by Swedish Match, who manufacture snus, along with a German tobacco wholesaler, contending that snus was a much safer product than cigarettes. However, the European Court of Justice rejected the challenge on the basis that at the time of the passing of the Directive, some scientists believed that such smokeless tobacco products could cause cancer and that they were also addictive.[26]

The regulations[14] define oral snuff as:

any product made wholly or partly of tobacco which is intended for oral use, unless it is intended to be smoked or chewed and is either in powder or particulate form or any combination of these forms, whether presented in sachet portions or porous sachets or in any other way, or presented in a form resembling a food product.

The distinction between what is legal and what is not is therefore somewhat arbitrary and based on whether the product is chewed or sucked. For example, a Danish smokeless tobacco product, Oliver Twist, which resembles moist oral snuff, is not banned because the packaging refers to chewing.

In 2005, in its first report on the application of the Tobacco Products Directive,[17] the Commission commented that separate sections were not included on tobacco products for oral use or hand-rolled tobacco because

the replies from the Member States [to a questionnaire on the Directive] *did not provide any new information as there was not enough new scientific information on ....tobacco products that may have the potential to reduce harm. However, the Commission continues to follow closely scientific developments as regards different aspects of tobacco products and their health effects, including carcinogenic, cardiotoxic and dependence producing effects, and will cover any findings in detail in the next report.*

The next report is due in 2007.

9.3.2 Chewing tobacco

As chewing tobacco is recognised within the definition of tobacco products used in the various regulations described in the section on combustible tobacco above, these regulations also apply to chewing tobacco. However, the voluntary agreement on control of additives was only signed by the manufacturers and importers of combustible tobacco products and so does not apply to smokeless tobacco products. Similar to combustible products, there are no regulatory controls on the raw tobacco itself. If therapeutic claims are made for the product, they must be licensed by the Medicines and Healthcare products Regulatory Authority (see below). Chewing tobacco packs are required to carry a health warning: 'This tobacco product can damage your health and is addictive'. Duty on chewing tobacco has been increasing in line with combustible tobacco but duty-paid markings are not required for chewing tobacco products.

9.3.2 Nasal snuff

Duty is not payable on nasal snuff products in the UK. However, advertising of nasal snuff products is not permitted and nasal snuff does fall within the definition of tobacco products within the EU product regulation directive[14] and so these regulations do apply. Nasal snuff is required to show the same health warning as chewing tobacco products.

9.4 New tobacco product formulations and potential reduced exposure products

Tobacco companies are free to bring new brands of cigarette, hand-rolling tobacco or chewing tobacco (other than oral snuff intended to be sucked) to the market without reporting to any regulatory authorities. Under UK regulations,[14] producers of cigarettes do, however, need to supply to the Secretary of State the names of brands to be produced and brands to be discontinued; tar, nicotine and carbon monoxide machine based yields of all these brands; and to provide product samples to be tested. Producers of every kind of tobacco product must also provide details of ingredients as described above, though this seems to have been ignored for new chewing tobacco products and combustible products to date. All new tobacco (oral or smoked) products must carry a health warning. There is no post marketing surveillance system for new tobacco products in place or required of manufacturers.

There is a proliferation of new tobacco products, some of which are not clearly conventional cigarettes or smokeless tobacco products, being launched in various countries. Those that resemble cigarettes are commonly referred to as potential reduced exposure products (PREPs). Some of these are conventional cigarettes containing tobacco with a reduced carcinogen content, or incorporating a filter device that removes a greater proportion of toxins from the cigarette. Others resemble cigarettes but are in fact novel devices that deliver nicotine by heating rather than burning tobacco or a tobacco derivative (see Chapter 5). It is not clear how these products will be regulated, or who will be responsible for their regulation, if and when such products are launched in the UK. The same applies for other novel tobacco-based products such as Nicogel, which is described as a tobacco gel, designed for topical use,[27] and is available for purchase in the UK.

9.5 Medicinal nicotine products

Nicotine replacement therapies (NRT) are regulated in the UK within the medicines regulatory framework, which is a comprehensive system first established in the Medicines Act of 1968. The system is currently implemented by the Medicines and Healthcare products Regulatory Agency (MHRA; formerly the Medicines Control Agency, MCA) with assistance from the Commission for Human Medicines. The MHRA has a UK-wide remit and is responsible to the Department of Health for regulating all medicines and medical devices by ensuring their efficacy and safety. The MHRA considers that any product claiming, or implying, that it can assist in the cessation of using nicotine products and smoking, is a medicine, and therefore comes within its regulatory framework.[28] Although weaknesses and failings within the MHRA and the medicines regulatory framework have been highlighted,[29] in comparison with tobacco products, regulatory oversight of NRT is very comprehensive.

9.5.1 Product

Nicotine gum was developed by Pharmacia in 1967, registered as a pharmaceutical and made available in Sweden in 1978, and introduced in the UK in 1981. The manufacturers demonstrated that NRT was effective at relieving withdrawal effects and increasing the chances of successful quitting, leading to the licensing of NRT as a medicine by the then MCA, for the relief of withdrawal symptoms as an aid to smoking cessation. As is standard practice, the potential risks and adverse effects of nicotine gum and the NRT products that followed (see Chapter 5) were assessed in relation to placebo, and as a consequence various contraindications and cautions were placed on the use of NRT.

Most of the regulation of smoking cessation medications is carried out by the Licensing Division within the MHRA, responsible for assessing and approving applications for product licences known as 'marketing authorisations' which set out conditions for use. Authorisations are required for new medicinal products or variations for medicines already licensed, such as new routes of administration, new formulations, or new uses. Hence a new nicotine product cannot be launched on the market without having been through the authorisation process. All applications are assessed by medical, pharmaceutical and scientific staff at the MHRA and often also involve independent expert advice outside of the MHRA through advisory bodies. Unlike cigarettes, licences are also needed for companies involved in all the stages of the manufacture and distribution of the product and this requires an inspection by the Medicines Inspectorate. The Licensing Division also has responsibility for authorising clinical trials of potential medicines. The MHRA also runs a post-marketing surveillance system for all medicines involving the reporting of adverse events (both through the pharmaceutical companies and health professionals through the Yellow Card reporting system), testing of medicines, inspection and enforcement.

Pharmaceutical companies are required to give a detailed description and composition of medicine products including the function and rationale for inclusion of each ingredient in the product, a detailed description of the manufacturing process and data on the stability of the product. This includes flavouring. The content of the active substances must be expressed quantitatively per dosage unit. The results of biological and toxicity testing are also required.

When the MHRA was formed in 2003 it adopted new objectives including 'making an effective contribution to public health'. Following representations from health organisations, in 2005 the Committee on Safety of Medicines (an advisory committee for the MHRA) set up a working group on NRT which examined the current evidence on efficacy and safety and made recommendations to relax the controls on the use of NRT products.[30] The key change was to accept that users of medicinal nicotine would otherwise be smoking tobacco, which is many times more harmful than medicinal nicotine. As a result, a number of changes to the licence were made, to extend the period of use to up to nine months, and to widen accessibility to the products to smokers with stable cardiovascular disease, pregnant smokers and to young smokers (from 12 years of age). In addition, two NRT products (Nicorette gum and Nicorette inhalator) were allowed to be used as part of a controlled strategy for quitting by cutting down on cigarette smoking in parallel with NRT use.

9.5.2 Place

There has been a progressive relaxation of initially strong restrictions on where and how NRT is available over the years since the products were introduced. Having first been excluded ('blacklisted') from availability on reimbursable NHS prescriptions, NRT products became reimbursible in April 2001. Three of the NRT products available on the UK market (gums, patches and lozenges) can now be sold on general sale (that is, through non-pharmacy retail outlets provided they are lockable and the product supplied in an unopened manufacturer's pack); the others (tablets, inhalators and nasal spray) are available only through pharmacies. The products on general sale are mainly sold in supermarket chains. The lack of NRT in the convenience market (in contrast to tobacco products which are available ubiquitously) is probably due to the stringent conditions regarding their accessibility as described above, their high prices (see below), and consequent low rates of sale relative to cigarettes.[31] NRT is most commonly packaged in weekly supplies but some smaller pack sizes (such as a two-day supply) are also available.

9.5.3 Promotion

Promotion of product information for drugs is under strict regulatory control via the Patient Information Quality Unit of the MHRA. Information to consumers is provided on the label or in patient information leaflets, which have to be approved in line with a European Council directive.[32] There are requirements to take patients' views into account in how this information is presented and received. Advertising to the public is permitted for medicines legally classified for pharmacy or general sale (prescription-only medicines may only be promoted to healthcare professionals). The advertising of all medicines is controlled through a combination of statutory measures (with criminal and civil sanctions) and self-regulation through codes of practice.

The restrictions applied by the MHRA on advertising are much stronger than those applied to other consumer products. For example, all parts of the advertising must comply with the particulars outlined in the Summary of Product Characteristics (SPC) and should encourage the rational use of the product by presenting it objectively and without exaggerating its properties. Advertising to the general public must contain a legible invitation to read carefully the instructions on the leaflet or packaging. The MHRA can require a viewing of the advertising before it is released.

9.5.4 Price

The price of NRT products to the NHS is controlled through the Pharmaceutical Price Regulation Scheme (PPRS), which determines the profit made by the pharmaceutical companies through the sale of their medicines to the NHS. When dispensed on prescription, NRT costs the smoker a relatively small fixed charge. Many smokers qualify for free prescriptions.

The price of NRT sold over the counter and in general sale is controlled by the pharmaceutical companies. For many years, NRT prices in the UK were broadly similar to the cost of regular smoking, though more recently the relative price of NRT has fallen. NRT is typically sold in packs containing a week's supply, which greatly adds to the purchase cost and may discourage purchase, particularly among lower income smokers. In the 2007 Budget, VAT on NRT was reduced from 17.5% to 5%, with effect from 1 July 2007 for one year.

9.5.5 New nicotine products

The recent government White Paper, *Choosing health*,[33] commented that the MHRA and pharmaceutical companies were discussing the development of new and innovative nicotine replacement therapies. However, innovation by the pharmaceutical companies appears unlikely while concerns persist that the MHRA would impose strong restrictions on products which more closely mimicked cigarettes and posed a greater chance of addiction than the current range of NRT products.

Nicotine products that do not contain tobacco do not necessarily have to be classified as a medicine. The MHRA decides whether any new nicotine product should be classified as a medicine on the basis of consumer perceptions, and claims being made by the manufacturers. A medicinal product is defined as:

(a) Any substance or combination of substances presented as having properties for treating or preventing disease in human beings;
(b) Any substance or combination of substances which may be used in or administered to human beings either with a view to restoring, correcting or modifying physiological functions by exerting a pharmacological, immunological or metabolic action, or to making a medical diagnosis.[28]

In the 1990s, Stoppers, lozenges containing nicotine, were sold unlicensed on general sale in the UK.[34] They were available for around 12 years before being withdrawn from the market.

9.6 Regulatory imbalance

The above discussion demonstrates the huge disparities and inconsistencies that exist between the tobacco and medicinal nicotine product regulations. Overall, combustible tobacco products are the least regulated and medicinal nicotine products are the most highly regulated. Given the huge differences in the proven or likely hazards of these products to individual and public health, this represents a substantial and illogical regulatory imbalance. The regulations in turn affect how the products are available to the public, with combustible tobacco products the most and medicinal nicotine products the least easily available, and also determine use. For example, the lack of access to NRT on a 24-hour basis in widely accessible outlets means that NRT purchases tend to be planned rather than made on impulse.

In addition to the imbalance between the regulations applying to the different nicotine products, there is a similar imbalance in the resources (human and financial) dedicated to the regulation of the products. In 2004/5, the MHRA had a budget of around £70m and 750 staff who can call on a number of advisory committees.[35] Whilst only a fraction of this resource is dedicated to NRT regulation, the available capacity is substantially higher than the budget and staff dedicated to tobacco regulation. Currently only one staff member in the UK (or English) Ministry of Health, a toxicologist, is dedicated to tobacco product regulatory work, although a proportion of tobacco policy team time is also used for product regulation. The former UK government scientific advisory committee (the Scientific Committee on Tobacco and Health – SCOTH) has not been reconstituted since its last term of office came to an end in 2004.

This situation contrasts starkly with other similar areas of relevance to public health. For example, in 2005–6, the Food Standards Agency had a staff of over 2,400, and a budget of over £135m.[36] Similarly the Health Protection Agency in 2006 had a budget of £235m and 3,000 employees.[37] Although the Wanless 2004 review[38] recommended in the Department of Health's review of arm's length bodies that 'Responsibilities should be assigned for:... the regulation of nicotine and tobacco', this did not happen.

9.7 Approaches to nicotine and tobacco product regulation in different countries

In developed countries NRT products are typically regulated within the country's medicines regulatory framework. To our knowledge, no country has co-regulated NRT and tobacco within the same regulatory system, although some, for example Canada, have the potential to do so as the same organisation regulates both products. The US has also made several attempts to have the same agency regulate the products. This section summarises the status of US tobacco product regulation, and then outlines a few examples from other countries where the regulatory structures or frameworks adopted differ from those in the UK.

9.7.1 Tobacco and nicotine product regulation in the United States

In the 1990s the Food and Drug Administration (FDA) in the US made an initial determination to regulate nicotine-containing cigarettes and smokeless tobacco products as combination drugs/devices within their jurisdiction. Following extensive public consultation, this culminated in a 'Final Assertion of Jurisdiction' being published in August 1996. As the FDA also regulated NRT products, there was a brief period of 'coregulation' of the nicotine market between 1996 and 2000. Legal challenges by the tobacco industry resulted in a decision by the Supreme Court in 2000 to overturn FDA's assertion of jurisdiction. The only way the FDA could then regulate tobacco products was by the enactment of new legislation by the Congress.

In early 2007, two identical bipartisan bills were introduced in the Senate and House of Representatives to grant the US FDA regulation of tobacco products by reinstating the 1996 FDA rule and also extending the provisions. The new provisions included the requirement from the FDA for detailed disclosure of ingredients, nicotine and harmful smoke constituents. They also empower the

FDA to require changes to current and future tobacco products to protect public health, and to regulate reduced harm products. FDA activity was to be funded through a fee on tobacco manufacturers allocated by market share.

These new bills have received support from various leading health organisations in the US and at least one tobacco company. However, there are concerns that the bar for reduced risk products entering the market is so high as to make it very difficult for the industry to introduce a reduced-risk product to the US market. For example, claims of reduced risk would need to be proven both at the individual and population level, and doing so could require comprehensive studies of non-smokers, ex-smokers, and smokers, of adults and children, carried out over several decades. This would be impossible in practice before bringing a product to market.

9.7.2 Canada tobacco and nicotine product regulatory framework

Including the regional offices, there are about 150 people working in the Tobacco Control Programme at Health Canada. The annual budget for tobacco control is about $60m Canadian dollars. The Canadian government has implemented legislation which requires comprehensive disclosure of tobacco products from tobacco manufacturers,[39] but has not yet regulated any aspect of product content or delivery. In Canada NRT products are also regulated by Health Canada, so there is the potential for co-regulation but this has not happened to date.

9.7.3 Regulation of snus in Sweden

In Sweden, snus is regulated as a food product under the Swedish Food Act. This is because the product is taken into the mouth and partly ingested. Regulating snus as a food means that only flavourings and ingredients that are accepted for food products are allowed in snus. As snus is regulated as a foodstuff the excise duties are comparatively lower than on cigarettes and other forms of tobacco.

9.7.4 Irish Office of Tobacco Control

In Ireland in 2002 a specific body, the Office of Tobacco Control (OTC), was set up to build capacity for tobacco control measures. The OTC manages all aspects of tobacco control. The implementation of the smoke-free policy in Ireland, which was the first national comprehensive smoke-free policy in the EU, heralded the way for several other countries to bring in similar policies following Ireland's example. The remit of the OTC also includes the potential to regulate the product, but this power has not yet been implemented due to a tobacco industry challenge on the Irish tobacco control legislation, and other tobacco control policies taking priority. Nicotine replacement therapies are regulated by the Irish Medicines Board.

9.8 Future regulatory options

The above sections have identified how nicotine and tobacco regulation in the UK and many other countries is currently weakest for the most harmful form of nicotine delivery, the cigarette. The inconsistencies and shortages in regulatory capacity for nicotine and tobacco product regulation have also been demonstrated. Restricting the choices available to smokers as to how they get their nicotine, and favouring smoked over non-combustible nicotine delivery systems, clearly works against public health. If the development of new and improved nicotine delivery systems is to be encouraged in the future, it is imperative to have a clear regulatory framework within which all nicotine products can be assessed in relation to their health impact. The aim of such a framework should be to reduce the health effects of tobacco use by minimising the use of nicotine-containing products overall but among regular users to maximise the use of safer nicotine products and minimise the use of combustible products. Having access to a comprehensive surveillance system would be critical in order to be able to respond quickly to any untoward changes in nicotine use. This section discusses the structural options for developing a coherent nicotine regulatory framework and steps that can be taken in the interim while such a framework is being developed.

9.8.1 Options for structural change

9.8.1.1 Regulate all nicotine products in the Medicines Regulatory Framework

In the 1970s the then Health Minister, Dr David Owen, proposed bringing cigarettes under the Medicines Act (a 'levelling up' option; see Chapter 11) but was moved from his post before the change was implemented. Although this proposition has not been formally raised since, and the definition of a medicine was subsequently changed, the fact that nicotine exerts a pharmacological action and modifies physiological functions could enable tobacco products to be regulated in this way. The MHRA, however, has not declared any interest in taking over tobacco regulation.

9.8.1.2 Take medicinal nicotine out of the Medicines Regulatory Framework

To correct the regulatory imbalance between medicinal nicotine and tobacco regulation, one solution would be to take clean nicotine products out of the Medicines Regulatory Framework and apply similar regulations to those currently used for tobacco (a 'levelling down' option; see Chapter 11). Given the weaknesses of, and piecemeal approach to, tobacco product regulation, there would be concerns about this among public health advocates. It is also likely to be an unattractive option to the pharmaceutical companies who would be left to compete with tobacco companies on the open market.

9.8.1.3 The establishment of a new nicotine and tobacco regulatory authority

A further option is to take tobacco and medicinal nicotine out of their existing regulatory frameworks and into a new structure. Current civil service structures mean staff turnover is frequent and greater capacity for tobacco regulation has not been forthcoming despite repeated calls over recent years. Creating a new institution to manage regulation has been the approach favoured in the UK for the regulation of drugs and food, and has been the preferred approach at least for tobacco regulation in other countries (such as Ireland). Establishing a single institution with a combined remit of tobacco and nicotine regulation would probably be the most efficient and coordinated way to enable a comprehensive approach to co-regulation of nicotine and tobacco products. A new institution would mean that a permanently staffed agency would be created with adequate authority to create an appropriate regulatory framework for tobacco and nicotine. Calls have been made for a nationalised tobacco industry or a state-owned monopoly for tobacco regulation.[40] These suggestions could be considered as the structure and the remit of the new authority is developed. As discussed in more detail in Chapter 12, the Royal College of Physicians believes that developing a new institution is the optimum approach to nicotine and tobacco products commensurate with the scale of the problem and the complexities of the regulatory responses needed.[41]

This issue was addressed in *Choosing health* in which the Government commented, without further justification of this statement, that 'we do not think that there is a case for setting up a brand new UK agency to regulate tobacco, as some have called for…'. The result is that the current piecemeal, inconsistent and irrational regulation of nicotine products continues to prevail.

9.8.2 Interim steps towards more coherent nicotine regulation

Whilst the establishment of a dedicated nicotine regulatory authority is likely to be the best means of delivering a rational and comprehensive nicotine regulatory framework that can maximise benefits to public health, there are steps that could be taken relatively easily in the interim to increase the accessibility of smokers to a wider range of medicinal nicotine products and to decrease accessibility to the most dangerous, smoked nicotine products.

The White Paper, *Choosing health*, outlined ways in which the MHRA and the pharmaceutical industry could work together to look at widening access and promoting use of NRT, including: 'developing new and innovative therapies', 'promotion of therapies through a wider choice of outlets', and 'encouraging retailers to allocate more space for stop smoking therapy products and space alongside cigarettes'. There has been little if any progress towards these objectives since the publication of the White Paper. More could be done within the existing regulatory system to encourage the improvement, development and promotion of medicinal nicotine products for smokers, by removing some of the real or perceived restrictions on nicotine product development, marketing and use. In particular, the current real or perceived likely restrictions on retail sale and display, and on marketing of nicotine products imposed by the MHRA could be relaxed substantially to increase accessibility to smokers, and to encourage the development and test marketing of more acceptable, more effective, and more affordable products. Current licence restrictions on long-term use of medicinal nicotine could be relaxed, and use as a means of cutting down on smoking and temporary abstinence from smoking encouraged more widely. Wider general availability of NRT in retail outlets could be encouraged, possibly in conjunction with a licensing system for tobacco retailers that requires them to provide alternative products could also be considered. All point of sale displays promoting cigarettes could be banned, and tobacco products required to be placed out of sight of customers. Consideration should also be given to any changes that can be made to reduce the harmfulness of cigarettes. One such current promising proposal is to reduce the ignition propensity of cigarettes, and a requirement to do this should be introduced without delay.

9.9　Conclusions

▸ Nicotine product regulation has developed in a largely reactive and piecemeal fashion over the years.

▸ Smoked tobacco products remained free from regulation for many years, and are now subject to minimal controls on content, delivery and safety.

▸ Some smokeless tobacco products are regulated very strictly (that is, they are prohibited) whilst others are subject to even less regulation than cigarettes.

▸ Medicinal nicotine products are regulated very strictly, as medicines.

▶ The lax regulation of most tobacco products affords considerable market freedom for tobacco companies to innovate and develop their products.

▶ The tight regulation of medicinal nicotine imposes very strict restrictions on new product development.

▶ Some newly launched tobacco products, including the PREPs, seem to lie completely outside of the current regulations.

▶ This clear and unjustifiable regulatory imbalance works against public health.

▶ UK government resources dedicated to tobacco product regulation are very small.

▶ History demonstrates that regulatory change can achieve substantial changes in consumption of different tobacco products.

▶ The regulation of nicotine products needs to be radically overhauled to encourage the use of less harmful products, and reduce the use of the more harmful sources of nicotine.

▶ Whilst some progress can be made in this regard through the existing regulatory systems, the establishment of a nicotine and tobacco regulatory authority is the preferred way of bringing comprehensive and rational controls on the nicotine product market that will minimise the harm caused by nicotine use.

References

1   Department of Health. *Smoking kills.* A White Paper on tobacco. London: The Stationery Office, 1999.

2   Joossens L, Raw M. The Tobacco Control Scale: a new scale to measure country activity. *Tob Control* 2006; 15: 247–53.

3   Doll R, Hill AB. Smoking and carcinoma of the lung. *BMJ* 1950;2:739–48.

4   Royal College of Physicians. *Smoking and health.* London: Pitman Medical Publishing, 1962.

5   Raw M, White P, McNeill A. *Clearing the air.* London: WHO Europe/BMA, 1990.

6   HM Customs and Excise, HM Treasury. *Tackling tobacco smuggling.* March, 2000. www.hm-treasury.gov.uk./media/6A1/17/433.pdf (accessed 7 August 2007).

7   HM Revenue and Customs. *Public Notice No. 477.* http://customs.hmrc.gov.uk/channelsPortalWebApp/channelsPortalWebApp.portal?_nfpb =true&_pageLabel=pageLibrary_ShowContent&id=HMCE_PROD1_026197&propertyTy pe=document

8   Department of Health and Social Security, Department of Health and Social Security Northern Ireland, Scottish Home and Health Department. *Fourth report of the Independent Scientific Committee on Smoking and Health.* London: HMSO, 1988.

9   HM Treasury, HM Revenue and Customs. *New responses to new challenges: reinforcing the Tackling Tobacco Smuggling Strategy.* March 2006. www.hm-treasury.gov.uk/budget/ budget_06/other_documents/bud_bud06_odtobacco.cfm (accessed 14 September 2007).

10   UK Trade Info. Tobacco Factsheet February 2007.
      www.uktradeinfo.co.uk/index.cfm?task=facttobac (accessed 14 September 2007).

11   HM Government. Tobacco Advertising and Promotion Act 2002.
      www.hmso.gov.uk/acts/acts2002/20020036.htm

12   WHO Framework Convention on Tobacco Control. Geneva, 2005.
      www.who.int/tobacco/framework/WHO_FCTC_english.pdf

13   The Tobacco Products (Manufacture, Presentation and Sale) (Safety) Regulations
      implementing EU Council Directive 2001/37/EC. www.opsi.gov.uk/si/si2002/20023041.htm

14   The European Parliament and of the Council of the European Union. Directive
      2001/37/EC of the European Parliament and of the Council of 5 June 2001 on the
      approximation of the laws, regulations and administrative provisions of the Member
      States concerning the manufacture, presentation and sale of tobacco products.
      Luxembourg, 2001. http://europa.eu.int/eur-
      lex/pri/en/oj/dat/2001/l_194/l_19420010718en00260034.odf

15   Tobacco Advisory Group of the Royal College of Physicians. *Nicotine addiction in Britain.*
      London: RCP, 2000.

16   National Cancer Institute. *Risks associated with smoking cigarettes with low machine-
      measured yields of tar and nicotine.* Smoking and Tobacco Control Monograph. Vol. 13.,
      Bethesda: US Department of Health and Human Services, National Institutes for Health,
      National Cancer Institute, NIH Publications, 2001.

17   European Commission. *Report from the Commission to the European Parliament, the
      Council and the European Economic and Social Committee.* First Report on the application
      of the Tobacco Products Directive. COM(2005) 339 final, July 2005. http://ec.europa.eu/
      health/ph_determinants/life_style/Tobacco/Documents/com_2005_339_en.pdf

18   Department of Health and Social Security, Department of Health and Social Security
      Northern Ireland, Scottish Home and Health Department. *Second report of the
      Independent Scientific Committee on Smoking and Health.* London: HMSO, 1979.

19   McNeill A, Godfrey F (eds). *Tobacco or health in the European Union. Past, present and
      future.* The ASPECT report. Brussels: European Commission; 2004.

20   House of Commons Hansard Ministerial Statements for 24 July 2006 pt 0166.
      www.publications.parliament.uk/pa/cm200506/cmhansrd/cm060724/wmstext/60724m0166.
      htm

21   King B, Borland R. What was 'light' and 'mild' is now 'smooth' and 'fine'; new labelling of
      Australian cigarettes. *Tob Control* 2005;14:214–15.

22   Wakefield M, Morley C, Horan JK, Cummings KM. The cigarette pack as image: new
      evidence from tobacco industry documents. *Tob Control* 2002;11(Suppl 1):i73–i80.

23   National Statistics and The Information Centre. *Statistics on Smoking: England 2006.*
      www.ic.nhs.uk/webfiles/publications/smokingeng2006/StatisticsOnSmoking300806_PDF.
      pdf

24   Council Directive 92/41/EEC of 15 May 1992 amending Directive 89/622/EEC on the
      approximation of the laws, regulations and administrative provisions of the Member
      States concerning the labelling of tobacco products

25   The Tobacco for Oral Use (Safety) Regulations 1992.
      www.opsi.gov.uk/si/si1992/Uksi_19923134_en_1.htm

26   European Court of Justice .Delivered on 10 September 2002(1). Case C-491/01.
      http//eur-lex.europa.eu/LexUriServ/site/en/oj/2005/c_045/c_04520050219en00070007.pdf
      (accessed14 September 2007).

27   See the manufacturer's website: www.nicogel-uk.com

28 MHRA. Guidance Note 8. A guide to what is a medicinal product. Revised March 2007. www.mhra.gov.uk/home/idcplg?IdcService=SS_GET_PAGE&nodeId=91

29 House of Commons Health Committee. *The influence of the pharmaceutical industry.* Fourth Report Session 2004-5. www.publications.parliament.uk/pa/cm200405/cmselect/cmhealth/42/42.pdf

30 Medicines and Healthcare products Regulatory Agency (MHRA). *Report of the Committee on safety of medicines working group on nicotine replacement therapy.* London: MHRA, 2005.

31 Judy Davis Marketing. *Nicotine replacement therapy. UK market review.* Report commissioned for ASH, May 2005. http://newash.org.uk/files/documents/ASH_429.pdf (accessed 14 September 2007).

32 EU directive on medicines Directive 2001/83/EC of the European Parliament and of the Council of 6 November 2001 on the Community code relating to medicinal products for human use. Official Journal L 311 , 28/11/2001 P. 0067 – 0128. http://europa.eu.int/eur-lex/lex/LexUriServ/LexUriServ.do?uri=CELEX:32001L0083:EN:HTML

33 Department of Health. *Choosing health: making healthier choices easier.* London: DH, 2004. www.dh.gov.uk/en/Publicationsandstatistics/Publications/PublicationsPolicyAndGuidance/DH_4094550

34 Foulds J, Russell MAH, Jarvis MJ, Feyerabend C. Nicotine absorption and dependence in unlicensed lozenges available over the counter. *Addiction* 1998;93(9):1427–31.

35 Medicines and Healthcare products Regulatory Agency. *Annual Report and Accounts 2004/5.* www.mhra.gov.uk (accessed 14 September 2007).

36 Food Standards Agency. *Putting the customer first. Annual report 2006.* www.food.gov.uk/multimedia/pdfs/annualreport0506.pdf

37 Health Protection Agency. *Annual Report and Accounts 2006.* www.hpa.org.uk/publications/2006/annual_report/pdf/Annual_Rpt_2006.pdf

38 Wanless D. *Securing good health for the whole population.* London, HM Treasury, 2004. www.hm-treasury.gov.uk./consultations_and_legislation/wanless/consult_wanless04_final.cfm

39 Health Canada Tobacco Act 1997. www.hc-sc.gc.ca/hl-vs/tobac-tabac/legislation/federal/tobac-tabac/index_e.html

40 Borland R. A strategy for controlling the marketing of tobacco products: a regulated market model. *Tob Control* 2003;12:374-82.

41 Royal College of Physicians. *Protecting smokers, saving lives. The case for a tobacco and nicotine regulatory authority.* Prepared by the Tobacco Advisory Group of the RCP. London: RCP, 2002.

10 Current nicotine product use and socioeconomic deprivation

10.1 Introduction

The use of tobacco products, and their consequent impact on health, varies substantially across society in relation to socioeconomic factors. Jarvis and Wardle identified what they called a general law of Western industrialised society that, 'Any marker of disadvantage that can be envisaged and measured, whether personal, material or cultural, is likely to have an independent association with cigarette smoking.'[1] They specify Western industrialised society as it is clear that their statement does not apply to countries in the earlier stages of the smoking epidemic. Their general law can also be extended to cover smoking cessation, in that any marker of disadvantage that can be measured is also likely to have an independent association with difficulty in giving up smoking. This chapter explores some of these effects in relation to smoked tobacco use in the United Kingdom.

10.2 Tobacco use among disadvantaged groups

There is a clear social gradient in smoking such that smoking rates are considerably higher among poorer people than among those who are better off. In 2005, 18% of men in professional and managerial occupations smoked, compared with 32% in routine and manual occupations (Fig 10.1).[2] Despite a reduction in the

189

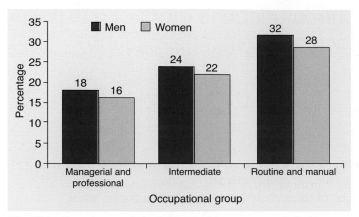

Fig 10.1 Prevalence of cigarette smoking among adults by sex and occupational group in Great Britain in 2005. Source: General Household Survey, 2005.[2] Reproduced under the terms of the Click-Use Licence.

overall prevalence of tobacco smoking in the UK over the past 30 years, smoking rates among lower income groups have remained significantly higher, maintaining the differential, and there has been little change at all in all occupational groups over the past decade (Fig 10.2).[2]

Furthermore, traditional measures of socioeconomic status tend to underplay the extent to which smoking has become concentrated in the poorest sections of society. Studies which have separated out the poorest in society, such as lone parents in receipt of social security benefits, have found smoking rates in excess

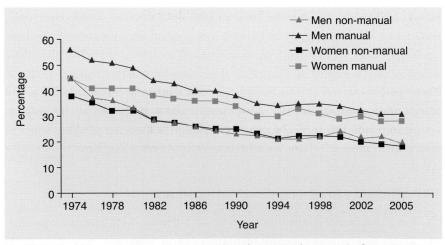

Fig 10.2 Smoking prevalence by sex and socioeconomic group in Great Britain in 1974–2005. Source: General Household Survey, 2005.[2] Reproduced under the terms of the Click-Use Licence.

of 75%.[3] Among the most affluent in the UK, smoking prevalence halved in the 30 years from 1973 to 2003, to only 16%. Among the most deprived, however, smoking rates have stayed the same: 70% were smokers in 1973 and 70% still smoked 30 years later in 2003.[1]

There are many other indicators of disadvantage that are linked to higher rates of smoking. For example, the odds of being a smoker are substantially increased in individuals without access to a car, who live in rented housing or in crowded accommodation, or who have no academic qualifications. Single mothers, including those who are divorced or separated, are also more likely to smoke. The following definable populations also have especially high smoking prevalence.

10.2.1 People with mental health problems

Smoking is significantly more prevalent among people with mental health problems than among the general population, and particularly among those with a diagnosis of a psychotic disorder. Smoking rates are as high as 80% among people with schizophrenia,[4] while over 70% of people with psychotic disorders who live in institutions smoke. Of these smokers, over half smoke more than 20 cigarettes per day.[5]

A national survey of psychiatric morbidity in over 8,000 people in the general population found that people with neurotic disorders such as depressive episodes, phobias or obsessive compulsive disorders were twice as likely as those with no neurotic disorder to smoke. Having more than one neurotic disorder was associated with heavy smoking.[6] Smoking has also been associated with adult attention deficit disorder,[7] eating disorders and substance abuse disorders.[8]

10.2.2 Prisoners

There were around 76,200 people in prison in England and Wales in 2005, and this number continues to rise.[9] Government estimates are that at least 80% of people in prison smoke.[10] The prison population is made up predominantly of young men, most of whom spend relatively short periods – weeks or months rather than years – in prison.

10.2.3 Black and minority ethnic groups

The Health Survey for England in 1999 found that smoking rates in many minority ethnic groups are lower than those in the rest of the population (Fig 10.3).[11] An exception was the smoking rate among Bangladeshi men, of whom 44% smoked.

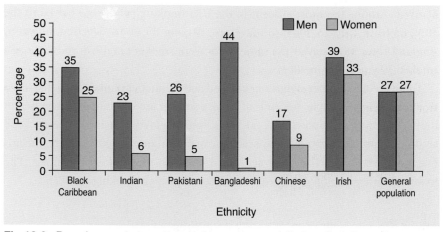

Fig 10.3 Prevalence of cigarette smoking among adults by ethnicity and sex.
Reproduced from Erens *et al* with permission of the Stationery Office.[11] Reproduced
under the terms of the Click-Use Licence.

The survey also found that the proportion of 'heavy smokers' (those smoking 20 or
more cigarettes per day) in all minority groups was lower than the proportion of
heavy smokers in the rest of the population (37%). The lower smoking rates among
some groups reflect cultural and religious differences, particularly among some
Muslim women. However, rates of chewing tobacco – almost unknown in the wider
population – are a health concern among some minority ethnic groups, particularly
in people from South Asia.[11]

The higher smoking rates among some minority ethnic groups may be attributable to the socioeconomic position of these groups. The Office for National
Statistics reported that minority ethnic groups were more likely than white groups
to live in low income households in 2000–01, although this varied by ethnic
group.[12] The highest rate was among Pakistanis and Bangladeshis of whom 60%
were living in low income households compared with 16% of the white population.
Nearly a half of black non-Caribbean households also lived on low incomes.

10.3 Consumption of tobacco and levels of nicotine dependence

Poorer smokers also tend to consume more tobacco than more affluent smokers,
which is in turn a strong indicator of higher levels of nicotine dependence.
Smokers in routine and manual groups consume on average 15 cigarettes per day,
compared with smokers in managerial and professional groups who consume
12 cigarettes per day.[2] There is also evidence that poorer smokers take in more
nicotine, and consequently more of the toxic emissions, from each cigarette

smoked – either by smoking cigarettes with a higher yield, by leaving a shorter stub, or by drawing harder on the cigarette. Data from the Health Survey for England show that the average level of saliva cotinine (a measure of nicotine intake) among the most deprived is 30% higher than among the most affluent (Fig 10.4).[1] The more deprived smokers are thus likely to experience greater harm from their smoking.

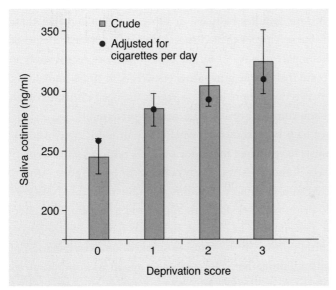

Fig 10.4 Saliva cotinine levels by deprivation in adult smokers in Britain in 2001. Reproduced from Jarvis *et al* under the terms of the Click-Use Licence.[1]

Indications of increased nicotine dependence among deprived groups in the UK are shown not just by quantitative differences in the amount smoked and nicotine metabolite levels, but also by factors such as time of first cigarette smoked and perceived difficulty of not smoking for a whole day.[2] Smokers in routine and manual occupations are more likely to report that they would find it difficult to go without smoking for a whole day (61% compared with 55% in intermediate and 50% in managerial and professional occupational groups) and were also more likely to have their first cigarette of the day within five minutes of waking up (18% in routine and manual groups, compared with only 10% in managerial and professional groups).[2] Since smoking-related disease is dose related, the fact that disadvantaged smokers tend to smoke more will directly increase their risk of mortality and morbidity. Also, since nicotine dependence is a critical indicator of ease of quitting smoking, this also indicates that disadvantaged smokers are also less likely to succeed in quitting smoking (see Section 10.7).

10.4 Costs of smoking

A smoking habit of 20 cigarettes per day at typical current UK retail prices costs about £1,900 per year. Poorer smokers spend a disproportionately large share of their income on cigarettes compared with more affluent smokers. In 2003, the poorest 10% of households spent 2.43% of their income on cigarettes each week, while the richest 10% of households spent 0.52% of their income.[13] In 1994, a study of the most deprived groups – including lone parents in receipt of state benefits – found that three out of four families had one or more smokers and that these smoking families spent one seventh of their disposable income on cigarettes, thus further exacerbating the impact of poverty on the family.[3] A more recent study in the north of England found that female heavy smokers were four times more likely, and male heavy smokers two times more likely, to report financial difficulties than non-smokers even after adjusting for age and quintile of deprivation.[14]

While price increases have been shown to be effective in reducing consumption, and some economists have argued that poor smokers respond more to price than do affluent smokers,[15] this does not always appear to be the case. Among the most heavily addicted smokers, many of whom are in the most disadvantaged groups, the response to price increases does not appear to be cessation but alternative strategies, such as switching to cheaper brands,[16] hand-rolled tobacco, smuggled cigarettes, cutting down on the number smoked, or a combination of these, rather than quitting altogether. However, as a result of compensatory smoking (maintaining nicotine intake by smoking fewer cigarettes more intensely) the health benefits achieved by cutting down the number of cigarettes smoked is likely to be minimised or even completely negated.[17]

10.5 Inequalities in the burden of ill health caused by smoking

One in two lifelong smokers is killed by smoking-related diseases.[18] In the UK, over 100,000 smokers die every year as a result of their habit,[18] and for every one who dies around another 20 are suffering from smoking-related diseases, many of whom will go on to die.[19,20] Smoking-related disease is a major cause of mortality and morbidity in disadvantaged groups, which have higher rates of smoking and lower rates of cessation. Additional factors such as poorer diet, earlier onset of smoking and higher levels of smoke intake (see above) exacerbate the risks.

Smoking is the biggest identified cause of inequalities in mortality between rich and poor in the UK.[21] It accounts for over half of the difference in risk of premature death across the range of socioeconomic status.[1] Death rates from tobacco are two to three times higher among disadvantaged social groups than among the better off.[22] Smokers in poorer social groups also tend to start to smoke at an earlier age: 48% of men and 40% of women in routine and manual occupations had become regular smokers by the age of 16 compared with 33% of men and 28% of women in managerial and professional occupations.[2]

Given the disproportionately high prevalence of smoking among people with serious mental health disorders, it is likely that a particularly high proportion of this population will die from a smoking-related disease. A 17-year prospective study in Finland, for example, found that having a mental disorder predicted an elevated risk of death from cardiovascular disease, respiratory disease and suicide.[23] The same study also demonstrated an association between schizophrenia and mortality from respiratory disease, which the authors suggest is also likely to be explained by smoking.

10.6 Disadvantage and smoking uptake

Children growing up in poverty and deprivation are more likely to live in homes where parents and/or siblings smoke,[2] so it is not surprising that they are themselves more likely to smoke. Parents and other family members are role models for the young and are a main source of primary socialisation. It has been shown that a significant reduction in the number of children taking up smoking is dependent on reducing smoking by adult role models.[24] Children are almost three times more likely to become regular smokers if both their parents smoke than if neither parent smokes.[25] Children growing up in an environment in which adult smoking prevalence is high are also more likely to experience adult smoking outside the home, which is also likely to increase the likelihood that they start smoking.[26]

In the national cohort of babies born in one week in 1958, the prevalence of cigarette smoking at age 16 ranged from 24% among those from the most affluent homes to 48% from the most deprived homes.[27] This gradient at age 16 was much sharper when cohort members were defined retrospectively by their own achieved social status seven years later at age 23, rather than by the characteristics of the parental household. In other words, the factors that led to a higher risk of smoking at age 16, such as poor school record and lower self-esteem, also led to an increased likelihood of subsequent downward social mobility. Furthermore, by their 30s, half of the better off young smokers had stopped smoking while three quarters of smokers in the lowest income group were still smoking.[27]

10.7 Smoking cessation in disadvantaged groups

The available evidence indicates that the desire to give up smoking is similar across social groups. Around two thirds of smokers, whatever their social groups, want to stop smoking.[28] Studies have also shown that around half of all smokers with mental health problems also want to quit.[5] The NHS Stop Smoking Services, set up over the past six years, represent a cost-effective and widely available source of help for smokers wishing to quit.[29] Recent research suggests that these NHS services have succeeded in attracting poorer smokers wishing to quit in areas of social deprivation. Nearly a third of all smokers in receipt of treatment services lived in the most disadvantaged quintile of areas while less than one in ten lived in the most advantaged quintile. An indicator of positive discrimination for each health authority area was calculated to quantify the extent to which the proportion of disadvantaged smokers being treated was greater than that in the population, and this ranged from just under 0% to 18%.[30] However, the overall contribution of the services to a fall in smoking prevalence is estimated to be only 0.1–0.3% per year.[31] Around a fifth of smokers in all socioeconomic classifications have used nicotine replacement therapy (NRT). More smokers in managerial and professional classes bought NRT over the counter, and more smokers in routine and manual occupations obtained their NRT on prescription, either paid for or free.[28]

However, even if poorer people are equally or more likely to make an attempt at quitting than more affluent people, higher levels of nicotine dependence make the likelihood of success lower. Rates of stopping smoking are three times lower among the least well off in society compared with the wealthiest (Fig 10.5).[2] The greater difficulties in quitting are exacerbated by fewer alternative diversionary or coping resources. Chances of success are reduced still further by a higher prevalence of peer group smoking and the greater likelihood of having a smoking partner.[32]

10.8 Exposure to second-hand smoke at work

The introduction of comprehensive smoke-free workplace legislation will help to reduce health inequalities, as levels of exposure to second-hand smoke at work otherwise tend to be higher among the less affluent workers in society. In Britain in 2004, only 4% of those in managerial and professional occupations worked in buildings with no restrictions on smoking, whereas 12% of those in intermediate occupations and 10% of those in routine and manual occupations worked in buildings with no restrictions on smoking (see Fig 10.6).[33]

Immediately before the implementation of smoke-free legislation in the UK began, 2.3 million UK workers were exposed to cigarette smoke throughout their workplace,[34] with hospitality workers particularly heavily exposed. About 10.4 million people worked in places where smoking was permitted in designated areas and were therefore also likely to have been exposed to tobacco smoke.[34]

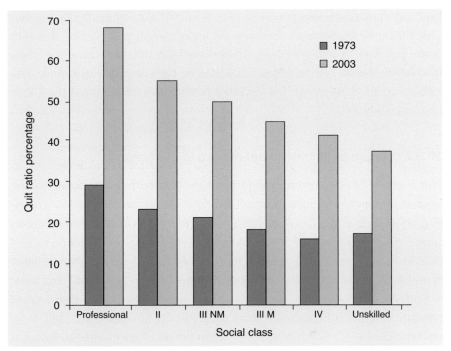

Fig 10.5 Smoking cessation by occupational social group in Great Britain in 1973–2003. I = Professional; II = managerial and technical intermediate; III NM = non-manual skilled; III M = manual skilled; IV = partly skilled; V = unskilled. Source: General Household Survey, 2005.[2] Reproduced under the terms of the Click-Use Licence.

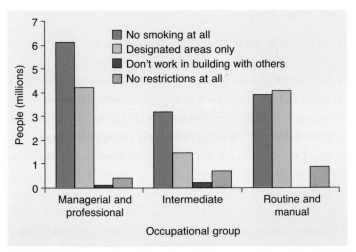

Fig 10.6 Exposure to smoke at work by occupational group in Great Britain in 2004. Source: Labour Force Survey and Office for National Statistics Omnibus Survey 2004.[33]

Exposure to tobacco smoke at work in the UK in 2003 was estimated to be responsible for the deaths of more than two employed people per working day (617 deaths per year), including 54 deaths in the hospitality industry each year.[35] These inequalities should now be largely eradicated by the comprehensive smoke-free legislation enforcing smoke-free enclosed public places throughout the UK in place since July 2007.

10.9 Exposure to second-hand smoke in the home

The majority of exposure to second-hand smoke occurs in the home, where legislation cannot be brought to bear. The health impacts are substantial, accounting for an estimated 2,700 deaths in adults aged 20–64 years, and 8,000 deaths among people aged over 65 in the UK in 2003.[35] Estimates of the proportion of children exposed to tobacco smoke at home vary from 32% to 42%.[36,37] Children of unskilled manual workers are substantially more likely to be exposed than those of professional people (54% and 18% exposed respectively),[37] and their level of exposure is typically higher since more people in the household are likely to be smokers (Fig 10.7).[24] Children who grow up in homes where parents smoke are at substantially increased risk of a range of childhood illnesses, including respiratory infection, middle ear disease and asthma exacerbation.[38] They are also approximately three times more likely to get lung cancer than the children of non-smokers, even if they do not take up the habit themselves as adults.[39]

10.10 Smoking and pregnancy

Smoking in pregnancy causes adverse outcomes, notably an increased risk of miscarriage, reduced birth weight and perinatal death.[40,41] Exposure to passive smoking during pregnancy is also an independent risk factor for low birth weight.[42–45] Babies exposed to their mother's tobacco smoke before birth grow up with reduced lung function.[46] Parental smoking is also a risk factor for sudden infant death syndrome (cot death).[47] The Infant Feeding Survey for 2000 found that 21% of non-smoking pregnant women were exposed to the smoke of someone else – usually a partner – who smoked in the home throughout their pregnancy. Living with a partner who smoked was highly correlated with low socioeconomic status.[48]

The Infant Feeding Survey also found that, in the UK in 2000, 20% of mothers smoked throughout pregnancy. Of mothers classed as 'never worked', 36% smoked throughout pregnancy compared with only 8% of mothers in managerial and professional occupations.[48] Younger mothers were also more likely to smoke

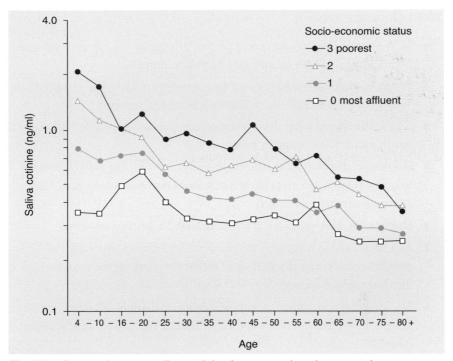

Fig 10.7 Geometric mean saliva cotinine in non-smokers by age and socioeconomic status, 1996–2003.[24]

throughout pregnancy than older mothers: 40% of mothers less than 20 years of age smoked throughout pregnancy compared with 13% of mothers aged 35 and over (Fig 10.8).[48]

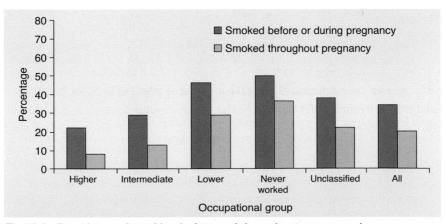

Fig 10.8 Prevalence of smoking before and throughout pregnancy by occupational group. Reproduced from Hamlyn *et al* with permission of the Stationery Office.[48] Reproduced under the terms of the Click-Use Licence.

10.11 Conclusions

▶ In Western industrial society, and particularly in the UK, cigarette smoking is strongly linked with socioeconomic disadvantage.

▶ People in disadvantaged socioeconomic groups are more likely to smoke, to smoke heavily, and to be more heavily addicted to smoking.

▶ This is also true of other disadvantaged groups, such as people with mental health problems or people in prison.

▶ Smokers in disadvantaged socioeconomic groups are just as likely to want to quit smoking, and more likely to use smoking cessation services, than relatively advantaged smokers.

▶ However, they are much less likely to succeed in quitting smoking.

▶ As a result, and in contrast to more advantaged groups, smoking prevalence has changed very little in recent years among the most disadvantaged in society.

▶ Smoking therefore causes even more death and disability in disadvantaged groups than in the rest of society, and is indeed the biggest cause of social inequalities in health.

▶ Children growing up in disadvantaged households are more likely to be exposed to cigarette smoke in the home, are more likely to start smoking, and to start when especially young.

▶ Disadvantaged smokers, and their children, therefore have the most to gain from harm reduction strategies for tobacco use.

References

1   Jarvis MJ, Wardle J. Social patterning of health behaviours: the case of cigarette smoking. In: Marmot M, Wilkinson R (eds), *Social determinants of health* 2nd edn. Oxford: Oxford University Press, 2005.

2   Office for National Statistics. *General Household Survey 2005*. London: ONS, 2006. www.statistics.gov.uk/ghs

3   Marsh A, McKay S. *Poor smokers*. London: Policy Studies Institute, 1994.

4   McNeill A. Smoking and mental health. A review of the literature. London: SmokeFree London, 2001.

5   Meltzer H, Gill B, Petticrew M, Hinds K. Office of Population, Censuses and Surveys (OPCS). *Surveys of psychiatric morbidity in Great Britain*. Report 1: The prevalence of psychiatric morbidity among adults living in private households. London: HMSO, 1995.

6   Health Development Agency. *Smoking and patients with mental health problems*. London: HDA, 2004.

7   Pomerleau OF, Downey KK, Stelson FW, Pomerleau CS. Cigarette smoking in adult patients diagnosed with attention deficit hyperactivity disorder. *J Subst Abuse* 1995;7:373–8.

8   Pomerleau CS, Ehrlich E, Tate JC *et al.* The female weight-control smoker: a profile. *J Subst Abuse* 1993;5:391–400.

9   Home Office. Prison population projections 2005–2010. 2005.

10 Department of Health. *Choosing health: making healthier choices easier.* London: DH, 2004.

11 Erens B, Primatesta P, Prior G. *The Health Survey for England: the health of minority ethnic groups, 1999.* London: The Stationery Office, 2001.

12 Office for National Statistics. *Census 2001: national report for England and Wales, 2004.* www.statistics.gov.uk/census/

13 Office for National Statistics. *Family spending.* Report of the 2002/03 Expenditure and Food Survey, 2004.

14 Edwards R, McElduff P, Harrison RA *et al.* Pleasure or pain? A profile of smokers in Northern England. *Public Health* 2006;120:760–8.

15 Townsend J, Roderick P, Cooper J. Cigarette smoking by socioeconomic group, sex, and age: effects of price, income, and health publicity. *BMJ* 1994;309:923–7.

16 Jarvis MJ. Supermarket cigarettes: the brands that dare not speak their name. *BMJ* 1998;316:929–31.

17 Russell MAH, Jarvis M, Iyer R, Feyerabend C. Relation of nicotine yield of cigarettes to blood nicotine concentrations in smokers. *BMJ* 1980;280:972–6.

18 Peto R, Lopez A, Boreham J *et al.* Mortality from smoking in developed countries, 2004. www.ctsu.ox.ac.uk

19 Cigarette smoking – attributable morbidity – United States, 2000. *MMWR Morb Mortal Wkly Rep* 2003;52;842–4. www.cdc.gov/mmwr/preview/mmwrhtml/mm5235a4.htm

20 Annual smoking – attributable mortality, years of potential life lost, and productivity losses – United States, 1997–2001. *MMWR Morb Mortal Wkly Rep* 2005;54;625–8. www.cdc.gov/mmwr/preview/mmwrhtml/mm5425a1.htm

21 Jha P, Peto R, Zatonski W *et al.* Social inequalities in male mortality, and in male mortality from smoking: indirect estimation from national death rates in England and Wales, Poland, and North America. *Lancet* 2006;368:367–70.

22 Acheson D. *Independent inquiry into inequalities in health.* London: The Stationery Office, 1998.

23 Joukamaa M, Heliövaara M, Knekt P *et al.* Mental disorders and cause-specific mortality. *Br J Psychiatry* 2001;179:498–502.

24 Royal College of Physicians. Exposure to passive smoking. In: *Going smoke-free. The medical case for clean air in the home, at work and in public places.* Report by the Tobacco Advisory Group of the Royal College of Physicians. London: RCP, 2005.

25 Office for National Statistics. *Teenage smoking attitudes in 1996.* London: The Stationery Office, 1997.

26 Jarvis MJ, Strachan D, Feyaraband C. Determinants of passive smoking in children in Edinburgh, Scotland. *Am J Pub Health* 1992;82:1225–9.

27 Ferri E (ed). *Life at 33.* The fifth follow-up of the National Child Development Study. London: National Children's Bureau, 1993.

28 Taylor T *et al. Smoking-related behaviour and attitudes.* London: ONS, 2005.

29 Godfrey C, Parrott S, Coleman T, Pound E. The cost-effectiveness of the English smoking treatment services: evidence from practice. *Addiction* 2005;100(Suppl 2):70–83.

30 Chesterman J, Judge K, Bauld, L, Ferguson J. How effective are the English smoking treatment services in reaching disadvantaged smokers? *Addiction* 2005;100(Suppl 2):36–45.

31 West R. *Stop Smoking Service quality and delivery indicators and targets.* Briefing for the Healthcare Commission, July 2004.

32 Jarvis MJ. Patterns and predictors of unaided smoking cessation in the general population. In: Bolliger CT, Fagerstrom KO (eds), *The tobacco epidemic,* vol 28. Basel: Karger 1997:151–164.

33 Action on Smoking and Health. *Smoking and health inequalities.* London: ASH, 2006.

34 Action on Smoking and Health. *Half the workforce still exposed to smoke.* Press release, 24 April 2006. http://www.ash.org.uk/html/press/060424.html

35 Jamrozik K. Estimate of deaths attributable to passive smoking among UK adults: database analysis. *BMJ* 2005;330:812.

36 Sproston K, Primatesta P. *The health of children and young people.* London: The Stationery Office, 2002.

37 Office for National Statistics. *General Household Survey 1997.* London: The Stationery Office, 1998.

38 US Department of Health and Human Services. *The health consequences of involuntary exposure to tobacco smoke.* Report of the Surgeon General. Atlanta, GA: US Department of Health and Human Services, Centers for Disease Control and Prevention, National Center for Chronic Disease Prevention and Health Promotion, Office on Smoking and Health, 2006.

39 Vineis P, Airoldi L, Veglia F *et al.* Environmental tobacco smoke and risk of respiratory cancer and chronic obstructive pulmonary disease in former smokers and never smokers in the EPIC prospective study. *BMJ* 2005;330:277

40 Eskenazi B, Bergmann JJ. Passive and active maternal smoking during pregnancy, as measured by serum cotinine and postnatal smoke exposure. I Effects on physical growth at age 5 years. *Am J Epidemiol* 1995;142:S10–18.

41 Windham GC, Hopkins B, Fenster L, Swan SH. Prenatal active or passive tobacco smoke exposure and the risk of preterm delivery or low birth weight. *Epidemiology* 2000;10:427–33.

42 Mainous AG, Hueston WJ. Passive smoke and low birth weight. Evidence of a threshold effect. *Arch Fam Med* 1994;3:875–8.

43 Misra DP, Nguyen RH. Environmental tobacco smoke and low birth weight: a hazard in the workplace? *Environ Health Perspect* 1999;107(Suppl 6):897–904.

44 Rebagliato M, Florey CV, Bolumar F. Exposure to environmental tobacco smoke in non-smoking pregnant women in relation to birth weight. *Am J Epidemiol* 1995;142:531–7.

45 Kharrazi M, DeLorenze GN, Kaufman FL *et al.* Environmental tobacco smoke and pregnancy outcome. *Epidemiology* 2004;15:660–70.

46 Gilliland FD, Berhane K, McConnell R *et al.* Maternal smoking during pregnancy, environmental tobacco smoke exposure and childhood lung function. *Thorax* 2000;55:271–6.

47 Anderson HR, Cook DG. Passive smoking and sudden infant death syndrome: review of the epidemiological evidence. *Thorax* 1997;52:1003–9.

48 Hamlyn B, Brooker S, Oleinkiova K, Wands S. *Infant Feeding Survey* 2000. London: The Stationery Office, 2000.

11 | Ethics, human rights and harm reduction for tobacco users

11.1 Background to the harm reduction approach
11.2 Components of the harm reduction argument
11.3 Harm reduction options for tobacco use
11.4 Approaches to rebalancing the nicotine product market
11.5 Conclusions

11.1 Background to the harm reduction approach

Smoking is dangerous to the health of the smoker and to others, including a fetus that is exposed to the smoke the smoker generates. Smoking is also highly addictive. Encouraging smokers to quit smoking, and developing better ways of helping them to do so, is a recognised public health priority. There is, however, a complementary approach, which is to reduce the harm caused by tobacco smoking by making effective but less hazardous substitute products available to the smoker.

An outline to this approach is as follows. The smoker who wishes to give up smoking faces many obstacles, particularly the psychological and physiological components of addiction. Different smokers use different routes to quitting, and may try several methods before finding one that works for them. However, many smokers continue to smoke in spite of a desire to quit, and even those who do not want to quit may wish to lessen the impact their smoking has on themselves and others. Since nicotine is the primary addictive constituent of tobacco smoke, the harm reduction approach for those who cannot otherwise quit smoking tobacco or who want to reduce the impact their smoking has on themselves and others is to substitute cigarettes with less hazardous alternatives. Even though smoking-related harms may be merely reduced rather than removed by this approach, many lives could also be saved and much morbidity prevented. For the purposes of discussion in this chapter we will consider the two broad groups of less hazardous alternatives discussed in Chapter 7 and Chapter 8: products based on medicinal nicotine, which we will assume to be the least hazardous alternative, and smokeless tobacco products, which we will assume to be more hazardous than medicinal nicotine, but much less hazardous than smoked tobacco products.

11.2 Components of the harm reduction argument

There are a number of objections to the harm reduction argument, just as there are for harm reduction approaches in relation to use of alcohol and psychoactive drugs. Doctors and public health experts are concerned with the empirical question: does making less hazardous products available and attractive to smokers actually, on balance, save lives, reduce tobacco-related morbidity and help smokers to quit smoking? There are many reasons why they may not: the existence of 'safer' products might cause some people who had previously been put off by the unitary 'smoking is dangerous' message to take up smoking or other tobacco product use; smokers might take longer to stop using tobacco with these products available than would otherwise be the case; the risks to health of the alternative products may be lower than cigarettes but are still a lot worse than not using nicotine products at all; and so on. But, as an empirical question, it is open to scientific study, and the harm reductionist and strict tobacco control theorist should be able to agree that such research is worth carrying out, and also that regulations which prevent this research are unhelpful.

A further objection to the harm reduction approach is an ethical one. The basis of this objection is that smoking is bad, as harm reducers allow, but promoting less harmful nicotine products as a substitute for smoking merely perpetuates the use of a drug in society. By this argument, harm reduction approaches dilute the message that smoking and indeed all nicotine or tobacco product use is or is potentially dangerous and undesirable. Furthermore, the consideration of smokeless tobacco products in this harm reduction strategy might allow the tobacco industry to promulgate an image of good corporate citizenship, while simultaneously continuing to market (and globalise) its highly dangerous cigarette products and promoting smokeless products as a vehicle to extend tobacco use. As former US Surgeon General, C Everett Koop, recently wrote:

> The problem with harm reduction approaches is that they may pose theoretical benefits to a few individuals, along with real and theoretical risks to many others. This was the experience with smokeless tobacco products in the United States, in which relatively few recalcitrant cigarette smokers may have switched from cigarettes to snuff, incurring theoretical (but as yet uncertain) disease reduction, while, at the same time, a new epidemic of smokeless tobacco was observed in young boys, who happened to be athletes.[1]

The force of this argument is both empirical and moral: the moral point is that the good intention of the harm reduction initiative induced a group of individuals otherwise rather unlikely to smoke to take up tobacco consumption. In essence, they had been 'corrupted'.

Arguments against harm reduction in the cases of heroin injecting, gambling or prostitution may focus on the immorality of the behaviour being regulated, whereas it is unlikely that many people think that smoking as such is immoral in present-day society. The moral force of these arguments in relation to tobacco use is directed instead at three parties: the tobacco industry in its ceaseless quest for new custom and profits; the harm reduction school of thought for abetting the industry in the sort of scenario Koop describes; and the wider community of people who see smoking as intrinsically morally unproblematic. In the cases of drug taking, smoking, alcohol consumption, gambling or prostitution, there are some who find these activities morally unproblematic so far as they represent conscious choices by consenting adults. There are also those who may find in these activities something to criticise morally, but would not see this as amounting to a case for legal prohibition or discouraging regulation.

Hence, critics of harm reductionist approaches to these activities see harm reduction as giving tacit support, in practice if not in theory, to these morally or legally permissive positions. And insofar as the critics of harm reduction see these kinds of activity as both wrong and to be legally prohibited, they reject harm reduction as failing to concede the moral wrongness of the activities involving harm. What critics of harm reduction are willing to accept, however, is that while taking this hard line will encourage some to desist from these activities completely, it will also leave others exposed to greater risk than they would be under a harm reduction approach, and that this may lead to greater aggregate harm than the harm reduction approach. They are willing to accept this on the basis that abstinence is morally preferable to 'moderated' consumption, and that moderated consumption is no better than unmoderated consumption. In a comparative example, they would see no difference between paying for sex in the context of a licensed brothel or in the context of unlicensed kerb crawling, even if the prostitutes and their clients may be safer in the former case. In the case of tobacco control, if you think that selling tobacco products involves a moral wrong, by deliberately exposing people to something that is harmful and addictive, then the moral wrong is hardly affected by the form in which the tobacco is sold.

It is not the view of the Royal College of Physicians that smoking is intrinsically immoral. Some consequences of smoking (such as non-consensual harm to others) may be morally wrong, and there are certainly morally difficult questions to ask about both the selling and consumption of highly addictive products which are dangerous to health. As with alcohol, however, these issues are sufficiently complex that no definitive position on the morality of smoking, insofar as it relates to the behaviour of individual smokers, has been reached. For the purposes of this report, therefore, we give no further consideration to the objections to harm reduction which focus on the morality or otherwise of smoking as such. Our focus is on a consideration of the harms to health posed by smoking and of proposals for policy to limit these harms.

11.3 Harm reduction options for tobacco use

In relation to tobacco use there are three basic harm reduction strategies available. The first is simply to maintain the status quo, in which efforts to prevent the harm caused by smoking are based entirely on strategies to achieve complete abstinence from nicotine use. In the second, nicotine products that represent an acceptable alternative to smoked tobacco (specifically cigarette smoking) to the smoker, but at substantially reduced relative risk, are made as available as cigarettes to adults. In the third, these alternative nicotine products are made relatively more available than cigarettes to adults. By 'more available' we include making them more affordable and more widely retailed. In the remainder of this chapter, the ethical arguments for and against each of these three options are summarised.

The ethical framework of analysis is a straightforward one. We consider in turn autonomy and individual rights, beneficence and paternalism, justice, and consequence-based arguments. Autonomy is the power each competent person has for making decisions about his or her own welfare and best interests. Beneficence is the moral obligation to benefit others. It is often argued that some public health or policy activity is justified on the grounds that it protects or promotes the good of those affected. Where this conflicts with the autonomous choices of competent people, or where they are not consulted (even if they might implicitly agree), imposition of the policy is termed 'paternalistic'. In reality, many public policy decisions are paternalistic in practice, and sometimes in principle, and such decisions require a moral justification in light of this. There has to be a good reason why the objective cannot be achieved through the autonomous choices of people themselves, and why the good of the objective outweighs the wrong done by overriding people's autonomy. Justice, for present purposes, concerns the fairness of the distribution of benefits and harms in society, such as health or costs of compliance with regulations. Autonomy, beneficence and paternalism, and justice-based arguments will generally be advanced in terms of rights (either legal, moral or human). Sometimes arguments are made which concentrate purely on weighing up the outcomes of different decisions and actions, without consideration of rights; these are consequentialist arguments.

Debates about public health issues such as tobacco regulation illustrate the ways that different kinds of moral principles interact. Public health concerns can override individual rights in specific circumstances in which an individual could cause harm to society as a whole. When there is low cost to the individual and high benefit to society, public health interests should prevail.[2] Public health ethicists have argued that it is not enough to speculate that there is *some* chance of adverse effects to society if an individual is allowed to exercise rights, it is necessary to show that there is *clear and convincing* evidence that the risk to society is substantial. In judging whether public health interests should override individual health interests, one needs to evaluate both the 'proportionality' and the 'probability' of the problems for public health.[3,4]

11.3.1 Model 1: the status quo

Autonomy and individual rights

Autonomy and individual rights are not given special priority in this model. No recognition is given to the right to exert autonomous choice over tobacco use by making choices from a range of acceptable and effective but substantially less hazardous products. It may be argued that promotion of autonomy does not imply making a wider range of tobacco products available. More choice need not mean more autonomy. Promotion of autonomy does mean realising the exercise of autonomy. Expanding the range of nicotine delivery products might actually impair autonomous choice by diverting choice from one harmful product to another, when what the smoker actually prefers is to give up smoking and to overcome the desire to smoke. Denying the smoker the means to realise this escape from addiction could be considered a constraint on autonomy. What is then at stake is whether alternative nicotine delivery products are an effective means of escaping tobacco addiction, and whether they help more people to stop smoking tobacco than more direct methods of quitting. Proponents of this model would argue that smokeless or novel nicotine delivery products should only be introduced into the market on the basis of good, clinical evidence that they are effective in aiding smoking cessation, and that they do not induce significant other harms from a public health point of view (such as being a gateway to active smoking). This would involve active regulation of such products on the model of medicinal products, as is indeed the case at present.

Beneficence/paternalism

It is considered wrong to release hazardous products into the market. Medicinal nicotine products are likely to be by far the least hazardous nicotine formulations, but if they are developed to become effective as cigarette substitutes they will almost certainly also be highly addictive. Smokeless tobacco is certainly addictive and more hazardous than medicinal nicotine. Aside from inherent risks to health arising from nicotine use, those who become addicted to medicinal nicotine or smokeless tobacco products might also, because of their addiction, switch to the very dangerous cigarette. In other words, alternative nicotine products could be a causal or relapse 'gateway' to cigarette smoking. This is a 'slippery slope' argument: if the step is taken to make alternative nicotine products more readily available, this will be problematic because of what it leads to (increased uptake of smoking or greater relapse). A similar argument is that some who have given up nicotine use altogether may now start to use nicotine products again, thereby increasing risk to those individuals. Also, there is concern that a product that is safer for individual users could result in a net loss for public health or society because of an increased numbers of users. All of these issues support the position of taking away the choice of individuals for their own good.

Justice

For our purposes, justice can be considered as fair treatment. In particular, one could ask: why is smoking tobacco available, but some smokeless tobacco products banned? Doesn't this discriminate unfairly against those whose preference is for smokeless tobacco? Yet this conclusion may not follow. There is arguably no right to consume tobacco, and consequently no right to consume it in any particular form. That is, no reasonable system of rights would allocate a specific right to people to smoke tobacco, anymore than it would allocate a specific right to consume caviar or watch television. At most, it might allocate a right to non-interference in one's private life, so far as what one is doing harms no one else. That smoking tobacco is not banned may be inconsistent with the ban on smokeless products, but the consequence of identifying this anomaly should be (as per the beneficence argument) to ban smoking as soon as may be possible, rather than relaxing restrictions on smokeless products. What justice requires instead is that vulnerable people should be protected from avoidable risks of addiction, and that people should not be hindered from promoting their own vital interests. Since smokers, as all of us, have a vital interest in their health, policy should not introduce measures which inhibit smokers' pursuit of that interest. But this needs to be balanced against the protection and promotion of the vital interests of all. Thus, for example, suppose that making smokeless tobacco products more easily available helped some smokers to quit smoking, but also delayed complete cessation in others, and induced some non-tobacco-users to take up tobacco use. What justice requires here turns on the empirical facts about what happens when smokeless, or indeed other nicotine products, are made available under different regulatory regimes.

Consequentialist perspectives

So far we have considered rights-based arguments. But, in practice, we need to consider the consequences of any policy proposed. To the extent that dangerous products are kept off the market, this should protect public health. The benefi- cence/paternalism arguments above can also be viewed as 'consequentialist' arguments; for example, the concern that lower risks to individuals distributed over more individuals could result in greater harm for society. The introduction of a 'safer' tobacco product could also create confusion as to the safety of any or all tobacco-based products. The adverse public health consequence might be that some current smokers continue to use harmful tobacco-based products (albeit in a different form) longer than they otherwise might, and some current non-smokers are initiated into tobacco use on the basis that they see the risk–benefit ratio of a smokeless tobacco product as being just favourable enough to make use attractive. Thus, it could be viewed as a strategic mistake, which might possibly be disastrous, to allow dilution of the central message.

11.3.2 Model 2: alternative nicotine products should made as available as cigarettes to adults

In this model, traditional tobacco products remain available on the same basis as now, with similar tobacco control policies in place so far as these relate to smoked tobacco, but regulation of alternative nicotine products (including smokeless tobacco and medicinal nicotine products) is relaxed to make them as easily available in terms of quantity, price, availability for sale and number of outlets, as smoked tobacco products. There are many ways to do this, from making alter- native nicotine products freely available, marketed in direct competition with cigarettes (for example, in low-cost, single day packs sold alongside cigarettes) at the most liberal end, to a position of restricting the distribution of tobacco products on the same basis as currently applies to nicotine replacement products. In other words, this model is about levelling the regulation of tobacco products for both smoked and smokeless products, be that levelling up or levelling down.

Autonomy and individual rights

The principles of autonomy and individual rights are that adults should have knowledge of and access to less hazardous forms of nicotine in case they want to choose to use them. If significantly less hazardous means exist to satisfy a nicotine addiction, honest information and availability are ways to respect individual rights. The current nicotine addict in particular may be viewed as someone who has the right to this information, in that the nicotine addict has a challenge and a great need to quit smoking.

This right to information is not extensive: it is a right not to be deceived or misled. Other things being equal, it is not ethically justifiable to promote a public health message (for example, that smokeless tobacco products are as, or almost as, dangerous as cigarettes) which is false merely because telling the truth is too complicated or might confuse people. However, we can also claim a stronger right: it is arguable that consumers have a right to know salient information about the products they wish to use, and about products which they may wish to use but are prevented from using on public health or product safety grounds. They have a right as citizens to know the reasons for decisions taken in the name of their best interests. While there may occasionally be circumstances under which certain kinds of information are state secrets for reasons of national security, or under which people are temporarily misled in order to prevent a panic, these kinds of reason hardly seem to apply to tobacco control.

For those who aren't satisfied that a citizen has a 'right to know', consider that the principle of individual autonomy supports the view that there is a human right to fair information relevant to healthcare and to products that would improve health. The Nuremberg Code (1949) and the United Nations Universal Declaration of Human Rights (1948) acknowledge a basic human right of autonomy. From legal perspectives, there are now expectations about patient autonomy and patient rights to be informed about, and to consent to, medical treatment.[5-7]

In addition to autonomy-based rights to accurate information about the comparative risks and benefits of smokeless and medicinal nicotine products, one can argue that the smoker has a right to access products which would enable him or her to give up smoking tobacco. Since smoking is an addiction, and cessation is difficult for many smokers, smokers face a conflict between their desire to smoke and their desire not to desire to smoke. The desire not to smoke has two components: a first order desire not to smoke, perhaps in response to social pressure, beliefs in the harmful effects of smoking on health or other similar reasons; and a second order desire not to want to smoke, that is, to overcome cravings. Reinforcing their autonomy means enabling them to allow this latter, second order desire to dominate the former, first order desire to smoke. Denying access to alternative nicotine products which could ease the transition from smoking to zero tobacco consumption inhibits their autonomy by making this reinforcement more difficult.[8]

Beneficence/paternalism

The beneficence/paternalism arguments raised above under Model 1 concerning possible 'slippery slopes' toward taking up tobacco smoking or making final tobacco-use cessation more remote remain salient, but they are now in conflict with the autonomy and individual rights arguments raised here. However, providing choice in nicotine products may benefit current smokers who wish to cease tobacco use in the context of a gradual 'weaning' off cigarettes, and will also benefit some of those affected by second-hand smoke by eliminating the source of that smoke in the case of smokers who switch to smokeless or medicinal nicotine products. This is not a paternalist argument save in the very weakest sense. On one reading, it is simply about extending choice, which is not paternalist in any sense. On a more plausible reading, it is about extending choice to promote or induce a move from a more to a less hazardous mode of consuming nicotine, with the hope that this will produce a move to cessation (which is mildly paternalist, but in an uncontroversial way). In other words, the beneficence/paternalism arguments may be consistent with the autonomy-based arguments used to support Model 2, provided there is good evidence that the balance of benefit in terms of smoking cessation is favourable with respect to any risk that smokeless tobacco products will encourage new smokers, or dissuade existing smokers, from achieving tobacco-use cessation.

Justice

From the point of view of consumers, it is unjust that they have ready access to particularly hazardous tobacco smoking products, but no legal access to some less hazardous smokeless products or to a more satisfactory medicinal product that could be used as a much less hazardous source of nicotine and arguably as a way to quit smoking entirely. Although, as argued under Model 1, there is no right to smoke (or otherwise consume) tobacco, there is a right to be protected from harms. Barring access to safer forms of the substance to which one is addicted could be seen as imposing avoidable harms if the tobacco user will be impelled by their addiction to use a more dangerous form in the absence of safer alternatives. From the point of view of manufacturers, it is unjust that one manufacturer should be allowed to market a very dangerous nicotine delivery system, while another manufacturer is not allowed to market, equally freely, much less dangerous nicotine delivery systems. This is especially the case when the basic justification for not marketing the product, as in the case of smokeless tobacco, is that they are dangerous. Manufacturer A, who sells cigarettes, is legally entitled to market his or her product; manufacturer B, who sells a form of smokeless tobacco, is not, even though the risks of B's product are less than those of A's. If the grounds for restraining B's trade are purely those of safety, then it seems A is being unfairly favoured over B on the grounds of a formal inconsistency in application of the principle underlying the restraint of B's trade.

Consequentialist perspectives

In the consequentialist perspectives above in support of maintaining the status quo in terms of regulation of medicinal and smokeless tobacco product restrictions, the focus was on the various bad things that might happen if restrictions, particularly on the use of smokeless tobacco products, were lifted. But there is a flaw in this argument. What various good things might fail to happen if the prohibitions are not lifted? It is an empirical question whether the number of novice tobacco users drawn into tobacco use by more widespread availability of smokeless tobacco or medicinal nicotine products, and the number of current tobacco users who defer cessation by switching to use of a lower hazard smokeless product, is or is not outweighed by the number of tobacco users who find direction cessation impossible but find cessation via switching from smoking to smokeless product or medicinal nicotine use possible. From a narrow consequentialist point of view, whereby the ethical justification of a policy is fixed by the net number of lives (or quality adjusted life years) saved, it matters what the facts are here about the relative risks of different kinds of products, likely transition rates from non-use to use of smokeless products, from smokeless product to smoking, from smoking to smokeless products, and so on. It is inadequate to simply speculate on what might happen, for example, with the availability of smokeless products.

When risks from a product used as a substitute for a more hazardous one are relatively small, the level of additional use needed to maintain a public health equilibrium (no changes in population-level problems) becomes very high. The risk to individuals from medicinal nicotine seems to be so low that it is not possible for use to increase enough to cause a net public health loss. That is, if risks from over-the-counter medicinal nicotine are, for example, less than 0.1% of those from cigarette smoking, then use would have to increase over 1,000 times more than any reduction in smoking achieved by their availability to cause an equal public health problem. For smokeless products such as snus, if the risk is as much as 10% of that of cigarette smoking, use would have to increase by 10 times the number of smokers who switched from smoking to smokeless use to generate an equal public health problem.[9,10] In relation to the theory that the availability of lower risk smokeless tobacco (or medicinal nicotine) may provide a causal gateway to cigarette smoking, there is some evidence that smokeless tobacco is a gateway for some to cigarettes, but there is also evidence that, for the majority of users, smokeless tobacco is not a gateway and may even help prevent smoking.[11-13] However, uptake of smokeless use by young people would still have to be very extensive to outweigh the public health benefits achieved in those smokers who switched to snus use. It is also important to appreciate that adolescent experimentation with cigarette smoking is the dominant gateway to a smoking habit in adulthood. A child at high risk of becoming a smoker would probably be better off using smokeless tobacco than cigarettes in this context, because the latter would greatly increase the odds of the child becoming an adult smoker, and whereas use of smokeless tobacco has a much reduced risk of leading to cigarette smoking. The above analysis takes no account of the additional benefits realised by switching from smoked to an alternative nicotine product on the 'third party costs' of smoking, such as the effects of smoking on non-smokers through passive smoking-related illness, fires, and so on.

11.3.3 Model 3: smokeless tobacco products should made more available than cigarettes to adults

Essentially, the arguments for this model combine the arguments for Model 2 with the general principle that the availability of smoking tobacco should be reduced, or the less radical alternative that smoking tobacco be driven down or out of the competitive market by greatly increasing the availability of smokeless tobacco and other nicotine delivery products.

Autonomy and individual rights

Current approaches to smoking tobacco do impose restrictions on its availability through restrictions on where, how and to whom it may be sold. These could be intensified in various ways: for instance, by raising the legal age at which one may purchase tobacco, requiring sellers to be licensed and issuing licenses very sparingly, or making tobacco a prescription-only product. This would clearly reduce autonomy in trying to block or restrict smokers' access to smoking tobacco rather than merely encouraging cessation through persuasion, education or the choice of safer alternatives. Price controls are arguably easier to defend than point of sale restrictions, since they operate through the autonomous choice to purchase smoking rather than alternative products. Thus, an expansion in the range of safer alternatives, which would tip the balance away from smoking tobacco and towards smokeless or nicotine replacement products, would be more consistent with autonomy and individual rights.

Beneficence/paternalism

Either approach (restriction of access to smoking tobacco or expansion of access to alternatives) can be seen as a beneficent/paternalist approach to alteration in smokers' behaviour. The argument here is essentially the same as that for Model 2. The difference is purely empirical: which approach has the better consequences in terms of the safety of tobacco users and in terms of eventual cessation, cultural change in the direction of disapproval of smoking, or other objectives of tobacco control?

Justice

It can be argued that restriction of access to smoking tobacco would be unfair, because it would impose disproportionate burdens on smokers who are addicted, though not if access restrictions leave smoked tobacco products available but less ubiquitously so (for example, through licensing a limited number of retail outlets). It may be that many of these individuals would find it easier to use alternative products, or to give up smoking, and this would be a good outcome. But not all smokers would be able to do this, and some might find the costs of alternatives excessive in relation to the costs of smoking tobacco (especially where the price of tobacco in black markets is taken into account) unless prices of alternatives were regulated with a view to promoting public health. Hence, from the point of view of justice, restriction of access to tobacco alone would need to be supplemented by expanded access to alternative products which are safer and no more expensive than smoking tobacco.

Consequentialist perspectives

The main issue here is whether adopting a restrictive access to smoking tobacco policy, with or without an expanded access to alternatives policy, would actually succeed in helping people avoid the harms of smoking. The obvious concern about a policy to restrict smoking tobacco would be that it might encourage illicit use of smoking tobacco and the associated illegal trade in smoked tobacco products. The obvious concern about an expanded access to alternatives policy would be that it might encourage continued use of products which remain harmful, rather than progression to cessation. The same arguments apply to Model 2, and the issue here is therefore an empirical one of what works best.

11.4 Approaches to rebalancing the nicotine product market

Changing the nicotine market, as outlined in sections 11.3.2 and 11.3.3 above, could be achieved through two different approaches: levelling down or levelling up. In the levelling down option, all alternative nicotine products that offer a clinically relevant reduction in risk could be deregulated to the same or greater extent than is currently the case for cigarettes and other smoked products, and market forces could then be left to determine whether existing and new less hazardous, more acceptable and more affordable nicotine products are promoted, brought to market and used by smokers. At the other extreme, it has been suggested that governments could take the tobacco industry into state control, either directly or through a third party organisation, and thus radically control promotion, distribution, branding and all other aspects of the tobacco market.[14] Both approaches are paternalistic, in that they attempt to influence consumer preferences in the direction of safer products, but they are also both consistent with the current regulation of smoking tobacco through taxation. The ethical issues of these options, or other approaches, go beyond those raised above, in that they mainly concern the ethics of regulating or otherwise controlling privately owned businesses through direct government action on the suppliers of the products, in addition to altering the behaviour of the consumers.

11.5 Conclusions

▸ Although stopping tobacco use is the ideal outcome for individual and public health, this is often difficult to achieve. Making a wider range of safer products available would be a harm reduction approach to tobacco control.

▸ Harm reduction approaches in public health are sometimes criticised for condoning the activity they are trying to make safer. The Royal College of Physicians takes no position on the morality of smoking. However, since smoking is dangerous to health, and is hard to give up, the College wants to see a range of effective methods to help smokers quit or to reduce the harm they sustain.

▸ The present status quo, in which cigarettes are freely available, medicinal nicotine products are available but under regulations that restrict availability and effectiveness, and some smokeless tobacco products are prohibited, denies smokers the right to choose safer nicotine products.

▸ Balancing the nicotine market, so that all nicotine products are equally available and comparably priced, would provide smokers with choice but would not encourage change from high risk to lower risk products.

▶ Rebalancing the market in favour of the safest nicotine products would provide choice, encourage safer nicotine use, and reduce morbidity and mortality.

▶ The ethical aspects of regulating alternatives to smoking tobacco are complex, and three positions can be defended: maintaining the status quo, making alternatives to smoking tobacco as easily available as smoking tobacco now is, or making them more easily available than smoking tobacco now is.

▶ Each alternative represents a balance of consumer rights, consumer protection, fairness and general policy considerations.

▶ The Royal College of Physicians favours an approach which would make smoking easier to give up, while discouraging people from starting to smoke in the first place, by making alternatives to smoking tobacco more widely available and by regulating smoking products more tightly through pricing, marketing controls and formal regulation of the production and sale of tobacco products. It believes that this offers the best balance of the ethical considerations surveyed in this chapter.

References

1   Koop CE. Tobacco: the public health disaster of the twentieth century. In: Boyle P, Gray N, Henningfield J, Seffrin J, Zatonski W (eds), *Tobacco: science, policy and public health.* Oxford: Oxford University Press, 2004:v–xvii.

2   Annas GJ. The impact of health policies on human rights: AIDS and TB control. In: Mann J, Gruskin S, Grodin M, Annas G (eds), *Health and human rights.* New York: Routledge Publishing, 1999.

3   International Federation of Red Cross and Red Crescent Societies, François-Xavier Bagnoud Center for Health and Human Rights. In Mann J, Gruskin S, Grodin M, Annas G (eds), *Health and human rights.* New York: Routledge Publishing, 1999.

4   Gostin L, Mann J. Toward the development of a human rights impact assessment for the formulation and evaluation of public health policies. In Mann J, Gruskin S, Grodin M, Annas G (eds), *Health and human rights.* New York: Routledge Publishing, 1999.

5   Nuremberg code: *Directives for human experimentation, 1949. Trials of war criminals before the Nuremberg military tribunals under Control Council Law No. 10, Vol. 2*: 181–2. Washington DC: US Government Printing Office.

6   Universal Declaration of Human Rights, 1948. Cited in Mann J, Gruskin S, Grodin M, Annas G (eds), *Health and human rights.* New York: Routledge Publishing, 1999.

7   Wear S. *Informed consent: patient autonomy and clinician beneficence within health care,* 2nd edn. Washington, DC: Georgetown University Press, 1998.

8   Elster J. *Strong feelings: emotion, addiction, and human behaviour.* Cambridge, MA: MIT Press, 1999.

9   Kozlowski LT, Strasser AA, Giovino GA, Erickson PA, Terza JV. Applying the risk/use equilibrium: use medicinal nicotine now for harm reduction. *Tob Control* 2001;10:201–3.

10   Levy DT, Mumford EA, Cummings KM *et al.* The relative risks of a low-nitrosamine smokeless tobacco product compared with smoking cigarettes: estimates of a panel of experts. *Cancer Epidemiol Biomarkers Prev* 2004;13:2035–42.

11   Hatsukami DK, Henningfield JE, Kotlyar M. Harm reduction approaches to reducing tobacco-related mortality. *Annu Rev Pub Health* 2004;25:377–95.

12   Kozlowski LT, O'Connor RJ. Apply federal research rules on deception to misleading health information: an example of smokeless tobacco and cigarettes. *Public Health Reports* 2003;118:187–92.

13   Ramström L. Snuff—an alternative nicotine delivery system. In: Ferrence R, Slade J, Room R, Pope M (eds), *Nicotine and public health*. Washington DC: American Public Health Association, 2001:159–74.

14   Callard C, Thompson D, Collishaw N. Transforming the tobacco market: why the supply of cigarettes should be transferred from for-profit corporations to non-profit enterprises with a public health mandate. *Tob Control* 2005;14:278–83.

12 Reducing the harm from nicotine use: implications for health policy and nicotine product regulation

12.1 Introduction

Smoking is powerfully addictive and kills half of all regular smokers.[1] As outlined in Chapter 1, having caused 100 million deaths in the 20th century,[2] and currently causing about five million deaths each year,[3] smoking is expected to result in a total of one billion deaths worldwide in the 21st century.[2] There are an estimated one billion smokers in the world, and this figure is expected to rise to 1.6 billion by 2025. Of all smoking products in current use, the cigarette is the most hazardous, the most addictive, and by far the most widely used.

Preventing people from ever starting to smoke is clearly the ideal means of preventing harm from smoking, but is only part of the solution for countries in which the smoking epidemic is already established. Preventing uptake of smoking will have minimal impact on deaths from smoking over the more immediate future because of the lead time of 20 years or more between starting smoking and the incidence of the main adverse effects on health. This means that the great majority of the global total of 150 million deaths from smoking expected in the next 20 years will occur in people who are smoking today. Preventing these deaths requires a focus on the needs of these individuals and helping as many as possible, as soon as possible, to stop smoking tobacco.

Current internationally endorsed tobacco control policies focus on both preventing the uptake of smoking and encouraging cessation by using taxation to increase the price of tobacco products, making workplaces and public places smoke-free, implementing effective health promotion campaigns, prohibiting tobacco advertising, providing cessation services, and other measures.[4,5] These policies are of proven effectiveness,[6] but even if fully implemented the magnitude of their effect on prevalence is modest. For example, in Massachusetts and California active tobacco control policy implementation achieved additional reductions in prevalence of around half a percentage point per year,[7,8] and in California this effect was not sustained in the longer term.[7] Even with full implementation of all recognised effective tobacco control policies it will take many years for a marked reduction in smoking prevalence, and in the morbidity and mortality that smoking causes, to be realised.

In the nine years since the publication of the White Paper, *Smoking kills*,[9] the UK government has taken a lead in implementing a wide range of tobacco control policies.[10] Cigarette prices in the United Kingdom are among the highest in Europe,[10] tobacco advertising is prohibited, smoking cessation services are available through the National Health Service to all smokers who want to quit, and comprehensive smoke-free policies have now been fully implemented. However, even if these measures succeed in maintaining or even increasing the recent rate of decline in UK smoking prevalence of about 0.4 percentage points per year for the longer term,[11] with 24% of British adults currently smoking it will take at least two decades for the prevalence of smoking in the UK to halve from current levels.[12] Even at that stage, there will be over five million smokers in the UK. These persistent smokers will include a high proportion who are heavily addicted and therefore find it especially difficult to quit despite many attempts to do so. There may also be a substantial number of 'hardcore' smokers who have no desire or intention to quit.[13] These smokers and their families, who tend to be predominantly from the most socioeconomically deprived sectors of society,[13,14] will bear a huge burden of potentially avoidable morbidity and mortality.

Current internationally recommended approaches to tobacco control are important, particularly in countries at an early stage of the smoking epidemic, and need to be maintained.[4,5] However, in countries such as the UK with a substantial population of established current smokers, tobacco control measures will not significantly reduce the shorter-term and medium-term burden of the morbidity and mortality caused by smoking, particularly among the most disadvantaged in society. In short, current tobacco control policy will fail to protect smokers who are unable to quit. New approaches are urgently needed to prevent death and disease in these people. Since they are not otherwise likely to quit smoking, the obvious alternative and public health priority is to find ways to reduce the harm caused by their habit.

12.2 The importance of nicotine in smoking behaviour

The evidence summarised in this report (Chapters 2, 3 and 4) demonstrates that nicotine in tobacco smoke is powerfully addictive. This addictiveness is probably enhanced by other components in the smoke, from physical stimuli arising from smoke inhalation, and possibly also by other related behavioural and environmental stimuli. Exposure to high levels of nicotine, and exposure from a relatively early age, may also result in neurodevelopmental changes that in turn influence the intensity of addiction. Once established, nicotine addiction is the principal underlying driver of sustained smoking behaviour.

12.3 The relative harm of different nicotine products

Nicotine is available in a wide range of products (see Chapter 5) in three broad categories: smoked tobacco, of which the cigarette is pre-eminent; medicinal nicotine, currently marketed as nicotine replacement therapy and intended for short-term use as a smoking cessation therapy; and smokeless tobacco products, of which oral tobacco is the most widely used.

Cigarettes and other smoked tobacco products are by far the most harmful of these nicotine sources (see Chapter 6). Cigarettes are designed to enhance the development and maintenance of addiction, and hence sustained use of the product. Cigarette smoke is harmful primarily because it delivers nicotine in conjunction with an extensive range of toxins and carcinogens.

By contrast, the safety record of medicinal nicotine products is extremely good (see Chapter 7). Although it is unlikely that medicinal nicotine (or any drug) is completely safe, there is no evidence that medicinal nicotine is carcinogenic, or that, in practice, its use increases appreciably the risk of acute cardiovascular events, or indeed that nicotine itself is responsible for much if any of the harm caused by cigarette smoking. To practical purposes, therefore, in comparison with smoking cigarettes, use of medicinal nicotine is extremely safe.

Smokeless tobacco products differ widely in their risk profile (see Chapter 8), largely in relation to the content of toxins in the tobacco. Of the more widely used products, Swedish oral tobacco (snus) appears to be the least hazardous. On current evidence it appears that snus use may increase the risk of cancer of the pancreas, and also of cardiovascular disease, but to a lesser extent than cigarette smoking. Snus use does not appear to be associated with an increased risk of oral cancer, but increased risks are well documented in association with use of some other smokeless products.

Unlike cigarette smoking, snus use does not cause lung cancer, chronic obstructive pulmonary disease or many other major adverse health effects associated with smoking. Thus, while undoubtedly more hazardous than medicinal nicotine, smokeless tobacco products, and particularly the low nitrosamine products, are substantially less hazardous than smoked tobacco.

12.4 The effectiveness of medicinal nicotine products as an alternative to smoking

Since first becoming available around 20 years ago, medicinal nicotine products have been licensed for use in the UK and in most other countries as cessation therapy. Although some national regulators now approve use for temporary abstinence or 'cutting down to quit', the great majority of medicinal nicotine has been used as a temporary substitute for nicotine from cigarette smoke during a quit attempt. Although use of medicinal nicotine increases the likelihood of quitting smoking by a factor of around 80%,[15] the overall effectiveness of medicinal nicotine is disappointing, since, in conjunction with best practice behavioural support, only about one in five smokers succeeds in quitting for six months or more.[16,17] Only a very small minority of smokers continue to use medicinal nicotine products in the longer term, and the risk of relapse to smoking increases substantially when use of medicinal nicotine ceases.[18] It is evident from the fact that relapse is so common among smokers quitting using medicinal nicotine products, and that long-term use of such products is so rare, that the available medicinal products are not strong substitutes for smoking. If these products were more effective they would be much more widely used, and probably more difficult to quit.

There are several likely reasons for this. Medicinal products are designed to minimise addiction potential by delivering nicotine more slowly and in lower doses than cigarettes. These characteristics also make them less effective as substitutes. Their packaging and pricing has tended to make them expensive at the point of sale to smokers, thus inhibiting impulse purchase or experimentation. This is likely to be especially so for less affluent smokers. They are marketed, packaged and promoted as cessation treatments, rather than as attractive and affordable competition to cigarettes. Although some are available for over-the-counter sale, their availability and display profile in retail outlets is substantially less than for cigarettes. Medicinal nicotine is also widely perceived by smokers to be harmful.[19]

12.5 Smokeless tobacco products as an alternative to smoking

Use of smokeless tobacco products predated cigarette smoking in many parts of the world, and continues to be widespread in many countries in which the smoking epidemic is still at an early stage. In countries at later stages of the smoking epidemic, the evidence on trends in use of smokeless tobacco, and on the gateway phenomenon by which smokeless products might encourage uptake of smoking or help existing smokers to stop smoking, is mixed.

In the United States, where smokeless tobacco has been used widely in the past, the prevalence of smokeless tobacco use has declined in recent years and is now low.[20] Smokeless products are not used widely in the USA as a smoking substitute by existing smokers, and the evidence relating to the concern that smokeless products might act as a 'gateway' into smoking is inconclusive (Chapter 8). In Europe, use of smokeless tobacco in the form of snus is prevalent in Sweden, where the evidence suggests that, in recent years, snus has been used among smokers predominantly as a substitute and/or cessation product. There has been relatively little gateway use by non-smokers into smoking. Snus is available in Norway but has not been used widely and does not appear to have had an appreciable effect on smoking behaviour. Use of smokeless tobacco elsewhere in the European Union, outside specific ethnic minority groups, is rare.

Thus, the Swedish data suggest that smokeless products, and particularly snus, may have the potential to act as effective substitutes for smoking, while experience elsewhere suggests that this effect may also be country or culture dependent, possibly arising from relative social acceptance of smokeless use, or from differences in perceived health risks among consumers. The Swedish data also indicate, however, that it may be useful to explore the possibility of harnessing some of the dose, delivery characteristics and associated stimuli provided by snus in the development of new medicinal products.

12.6 Harm reduction strategies

Harm reduction strategies are pragmatic approaches to reducing the harm arising from a hazardous behaviour. In broad terms, harm reduction is fundamentally embodied in the application of common sense and the widespread regulatory controls intended to minimise the inevitable hazards of everyday life. In relation to tobacco smoke exposure, legislation prohibiting smoking in enclosed public and workplaces is an example of a harm reduction strategy intended to protect workers from the adverse effects of passive smoke exposure.

Harm reduction strategies are used widely in public health to reduce the harm caused, for example, by illicit drug use and a range of other behaviours. The rationale behind their application in this context is that while the best option for individuals and/or society would be to avoid harmful behaviours completely, the next best option, if avoidance is not a practical or realistic option, is to minimise the harm caused by the behaviour.

Cigarette smoking is a highly hazardous and addictive behaviour, and complete abstinence from smoking is the obvious best option for health. However, as argued in this report, this option is not realistically achievable in the short-term or medium-term future for a substantial proportion of smokers in populations in which smoking is already established. In this context, since tobacco smoking is driven primarily by addiction to nicotine, but the harm from smoking is caused by other smoke constituents, the rational next-best option is to reduce the harm arising from nicotine use by providing it in a form that does not involve inhaling smoke. The alternative sources of nicotine offered need to be acceptable to smokers as substitutes for cigarettes, and available at a price and with marketing and health messages necessary to encourage smokers to substitute them for tobacco smoking. They also need to be substantially less hazardous than smoking.

Effective harm reduction strategies have not, to date, been applied to any significant extent to tobacco smoking. Controls based on machine-measured cigarette yields have been ineffective. Attempts to restrict the toxicity of conventional cigarettes by altering the content of tobacco or other measures used in potential reduced exposure products (PREPs) (Chapter 9) have not been shown to deliver real reductions in risk, and will not have an appreciable effect on the risk of smoking unless they result in radical reductions in exposure to toxins and carcinogens. The option of providing nicotine without smoke, intended as a long-term or, if necessary, lifelong substitute for regular smoking, has not been explored. Medicinal nicotine products have been developed and used almost exclusively as temporary cessation aids. Use of smokeless tobacco products, although substantially less hazardous than smoking, is currently actively discouraged, and in relation to some products in the European Union, prohibited.

The fundamental argument of this report is that this current situation is perverse, unjust and acts against the rights and best interests of smokers and the public health. Harm reduction has the potential to play a major part in preventing death and disability in the millions of people who currently smoke and who, in the context of exposure to currently available drivers and supports to cessation, either cannot or will not otherwise quit smoking. These smokers have a right to be able to obtain and choose from a range of safer nicotine products (Chapter 11), and they have a right to accurate and unbiased information to guide that choice. In a recent study, it was estimated that if a product such as snus were introduced into the US market and promoted with a warning stating,

'This product is addictive and may increase your risk of disease. This product is substantially less harmful than cigarettes, but abstaining from tobacco use altogether is the safest course of action', the prevalence of smoking in the USA would decline by between 1.3 and 3.1 percentage points over five years.[21] That is an annual decline of between 0.25 and 0.6 percentage points per year, or approximately 0.4 percentage points per year; sufficient to double the recent rate of decline in the UK.[11] A switch of only 0.4% of the population of smokers in the UK each year from smoking to less harmful nicotine sources – a conservative target – would save around 25,000 lives in only 10 years.[22]

There are, however, a number of obstacles and barriers to the development of an effective harm reduction strategy for tobacco smoking. These include the moral concerns of health professionals and others outlined in Chapter 11, and reluctance by governments to engage in a difficult and controversial shift in policy. However, one of the main obstacles to progress is the current system of regulation that applies to nicotine products.

12.7 Current nicotine product regulation and health

Regulation of nicotine products varies between countries, but, as outlined in Chapter 9, has tended to evolve in piecemeal reactive legislation for tobacco products, while medicinal nicotine has been subject to the much stricter regulatory frameworks that most countries apply to medicines.

Cigarettes and other smoked tobacco products have enjoyed almost complete freedom from regulatory control in many countries for many years, and although now subject to a growing range of restrictions on machine-measured nicotine and tar yields, and on advertising and health warnings, smoked tobacco products remain remarkably free from any regulation likely to have an appreciable effect on the harm caused to the consumer. Indeed, a major reason why tobacco products have remained exempt from consumer protection regulation for so long is that their status as a legal product is so inconsistent with their risk profile that the proportionate application of the regulatory systems that control other consumer products would result in their immediate withdrawal from sale. As a consequence of this freedom from effective health, safety and consumer protection regulation, the most dangerous and addictive nicotine products remain regulated to a minimal degree and in extreme disproportion to their hazard, and are freely available and widely used. Under the current regulatory system, tobacco companies remain free to develop or modify, and bring to market, new smoked tobacco products, PREPs, and other tobacco derivatives with little by way of regulatory control.

Medicinal nicotine products, in contrast, are strongly regulated as drugs. This has resulted in the development and marketing of nicotine products that have low addiction potential, deliver low nicotine doses, and are promoted and marketed for short-term use as cessation aids. While this may be justified in the context of smoking cessation objectives, these characteristics are almost entirely the opposite of those needed if medicinal nicotine is to prove an effective and acceptable cigarette substitute. Thus the safest available nicotine products are currently subject to the highest level of regulation, and the most dangerous to the least. The restrictions that these regulations impose on the likely uptake and availability of medicinal nicotine products may have acted as a major disincentive to innovation and competition in the medicinal nicotine market.

In terms of relative hazard, smokeless tobacco products vary, but all are less hazardous than smoking and some are especially so. The regulations imposed on these products in the UK and most EU countries are entirely inconsistent, both within the range of smokeless tobacco products (since the least hazardous are the most regulated) and also in relation to medicinal or smoked nicotine. Some products, such as chewing tobacco or nasal snuff, are available and subject to minimal regulatory controls, yet are more hazardous by comparison with other smokeless products. Others, such as snus, appear to be the least hazardous yet are widely prohibited. In view of the low hazard associated with low-toxin oral products such as snus, and evidence of the potential of these products as smoking substitutes, their prohibition in the context of free availability of other smokeless and smoked tobacco products is irrational. The potential for existing or new smokeless products to provide an alternative source of nicotine for smokers, to complement medicinal products, merits further investigation. While rationalisation of smokeless product regulation by extending the prohibitions that apply to some products across the full range of smokeless tobacco would resolve the present inconsistency, the danger is that if smokeless products can help some smokers to quit who would not otherwise do so, without other marked adverse effects on public health, this potential benefit will be lost. At the other extreme, lifting the prohibition on low hazard products and in effect further extending the free hand currently enjoyed and exploited by the tobacco industry to a new product area is potentially disastrous. The market in smokeless tobacco, as for all nicotine products, needs consistent and rational monitoring and control.

12.8 How should nicotine products be regulated to improve public health?

Nicotine products need to be regulated pragmatically to reduce the overall harm caused by nicotine dependence and use. This will involve trying to reduce overall levels of nicotine product use, and also reducing the proportion of nicotine users who use the most hazardous nicotine delivery devices, particularly cigarettes. Achieving this will require radical overhaul. The current nicotine regulatory framework needs to be changed so that it encourages as many smokers as possible to quit smoking and all nicotine use completely, and encourages those who cannot quit to switch to a safer source of nicotine, while minimising use by people who would not otherwise have used nicotine products. It should not encourage new users to start using any nicotine product. The framework also needs to encourage innovation, development and ultimately the use of new medicinal nicotine products at the less hazardous end of the spectrum, and to demand significant reductions in hazard, availability and affordability of products at the smoked tobacco extreme.

Overall, the framework should ensure that the market forces of affordability, promotion and availability to the consumer apply strongly and in direct inverse relation to the hazard of the product, thus creating the most favourable market environment for the least hazardous products, and discouraging the use of more hazardous products. The anomalies that inhibit new product development, in particular rapid delivery, user-friendly medicinal products and more widespread use of existing low-hazard products, need to be removed. The regulatory system for nicotine products must also be able to respond to and mitigate potentially counterproductive trends in marketing or use of nicotine products, and must therefore have access to regular, frequent surveillance data on patterns of use, marketing strategies and health claims by manufacturers. It must also ensure that alternative nicotine products, be they medicinal or tobacco-based, are marketed with appropriate health information.

Nicotine product regulation should also evolve so that smoked tobacco products are subject to progressively increased restrictions on availability and marketing, with the longer-term objective of minimising and, in due course, eradicating the use of smoked tobacco. A coherent and planned strategy to achieve these objectives would need to be pursued by government, the medicines regulators, the industry, and the medical and smoking cessation community.

The ideal outcome would be a coherent and consistent nicotine regulatory system that takes responsibility for all nicotine products. There are many potential ways of achieving this outcome, such as bringing all nicotine products under the responsibility of an existing agency (such as food or drug regulation agencies), or by creating an overarching coordinating body to oversee and complement the activities of the agencies currently involved in regulating nicotine products. However, our firm conclusion is that the special problems of regulating nicotine products, and providing market surveillance, clear consumer information and monitoring industry activity, requires special skills and concentrated expertise that would be best established in a dedicated nicotine product regulatory authority. Some of the initial and continuing functions of the proposed authority are outlined in Box 12.1.

While working towards a nicotine regulatory authority, we believe that significant changes to the way that nicotine is regulated by the Medicines and Healthcare products Regulation Agency (MHRA) could easily be made in the more immediate future. This would enable the 'new and innovative therapies' envisaged in the White Paper, *Choosing health*, to be developed and promoted to smokers.[23] Some relaxation of the restrictions on use of medicinal nicotine have already been made by the MHRA, which now accepts that users of nicotine products would otherwise be smoking, and that the risks of use of nicotine products should be assessed in that context. However, we would like to see the MHRA extending this harm reduction model to other aspects of medicinal nicotine product development and use, in order to promote the availability of fast-acting, user-friendly products that are easily available, affordable, and likely (if effective) to be used for the longer term. Controls on advertising of medicinal nicotine products could be relaxed to allow manufacturers to promote medicinal products on a harm reduction platform, thus also helping to break down public misconceptions of the risks of nicotine use. Simplification of the licensing process and other measures that encourage competition in the medicinal nicotine market to generate better and more affordable products are likely to be beneficial to public health. The recent reduction of taxation on medicinal nicotine products by the UK government is a welcome example of a simple innovation that is likely to encourage greater use.

Implementing a strong harm reduction strategy for smokers would give smokers the choice, currently not available to them, to use products that are tens of times if not hundreds of times less hazardous than cigarettes, so saving lives and significantly reducing social inequality in health. It would also support the introduction of smoke-free legislation, help to denormalise smoking, significantly reduce exposure to passive smoke in the home (where children in particular are most heavily exposed) and help to reduce the number of smoking-related fires in both domestic and commercial settings. Such a strategy would be a market-based, low-cost public health intervention.

12.9 The consequences of a failure to act

The inevitable consequence of failing to address the special problems of nicotine product regulation will be the perpetuation of current smoking among millions

Box 12.1 Suggested roles and functions of a national nicotine regulatory authority.

Functions at initiation

▶ Baseline measurement of all current nicotine product use

▶ Ensure full implementation of conventional tobacco control policies

▶ Permissive licensing of medicinal nicotine products for use as smoking substitutes

▶ Substantial relaxation of restrictions on marketing and sale of medicinal nicotine products

▶ Removal of tax on medicinal nicotine products

▶ Communication of objective health risk information for nicotine products and promotion of harm reduction principles to smokers and the public

▶ Establishment of ground rules for monitoring the use of health messages in promoting the use of lower hazard nicotine products as substitutes for smoking

▶ Imposition of generic packaging for all tobacco products

▶ Prohibition of retail display of smoked tobacco products

▶ Strong graphic health warnings on smoked tobacco products

▶ Setting of tax and consequently retail price of all nicotine products in relation to their likely relative risk to health

▶ Prohibit all sale of nicotine products to individuals aged under 18

▶ Introduce licensing of retailers of all smoked tobacco products

▶ Assume responsibility for overseeing nicotine product delivery and toxicity monitoring

▶ Mandate the introduction of reduced ignition propensity cigarettes

▶ Take expert advice on how current restrictions on smokeless tobacco could be reformed to public health benefit

Continuing functions

▶ Regular monitoring of trends in nicotine product use, promotion and availability

▶ Monitoring impact of licensing and marketing relaxation on medicinal nicotine use, and revision as necessary to promote public health

▶ Progressive increases in tax on the most hazardous products

▶ Continued promotion of health information on different nicotine products and development and monitoring of mass communication strategies to prevent uptake, promote cessation, and reduce harm

▶ Progressive reduction in retail licences for smoked tobacco products

▶ Monitoring and policing of illicit and underage tobacco and nicotine trade

▶ Work with the commercial sector to promote competition and innovation in the medicinal nicotine market

▶ Monitoring and prevention of smoked product placement and new methods of marketing (eg internet, viral marketing)

▶ Act on expert advice to set framework for licensing of low-hazard smokeless products and possible test marketing

▶ Progressively incentivise minority, high risk smokeless tobacco users to quit or else migrate to safer products

▶ Identify and respond to new developments or threats to health from new or existing product development or promotion

▶ Control of expenditure on tobacco control interventions to ensure evidence-based and cost-effective interventions are used

▶ Support nicotine regulation and tobacco control approaches in resource-poor countries

of people, and a consequent continued epidemic of death and disability caused by tobacco smoking. Specifically, cigarettes and other smoked tobacco products will continue to be marketed with minimal restriction on their safety or content; smoking products alleged to present lower risks to the smoker, such as PREPs, will be freely marketed and their health claims unsubstantiated; the medicinal nicotine market will continue to focus on low-addiction, low-dose, low-effectiveness products while also stifling competition and innovation; the current piecemeal and inconsistent regulation of smokeless products will continue, preventing smokers in most EU countries from choosing to use a significantly less hazardous tobacco product than cigarettes; and the majority of smokers alive today will continue to smoke, and half will die as a result.

12.10 Conclusions

▶ Most of the deaths and disease caused by smoking in the near- and medium-term future will occur in people who are smoking now.

▶ Current conventional preventive measures focus entirely on preventing uptake of smoking and helping smokers to quit smoking.

▶ This approach will be ineffective for the millions of smokers who, despite best efforts to persuade and help them to quit, will carry on smoking. Half of these smokers, representing millions in the United Kingdom alone, will die as a result.

▶ This burden of mortality and morbidity will markedly exacerbate social inequality in health.

▶ Tobacco control policy needs to be radically extended to address the needs of these smokers, by implementing effective harm reduction strategies.

▶ Harm reduction in smoking can be achieved by providing smokers with safer sources of nicotine that are acceptable and effective cigarette substitutes.

▶ There is a moral and ethical duty to provide these products to addicted smokers.

▶ Current systems of regulation of nicotine products inhibit the development of innovative medicinal nicotine substitutes for cigarettes, and perpetuate the use of the most dangerous nicotine products. This is unjust, irrational and immoral.

▶ Nicotine product regulation must, therefore, be reformed.

▶ The unprecedented and unjustifiable market freedoms enjoyed by manufacturers of cigarettes and other smoked tobacco products must end.

▶ The development of new, more effective, more acceptable and user-friendly medicinal nicotine substitutes for smoking needs to be encouraged.

▶ A major step towards an effective harm reduction strategy could be taken by the implementation of simple changes to the regulation, promotion and taxation of medicinal nicotine.

▶ Low nitrosamine smokeless tobacco products may have a positive role to play in a coordinated and regulated harm reduction strategy which maximises public health benefit and protects against commercial market exploitation.

▶ The regulation of nicotine products, be they medicinal or tobacco-based, thus needs radical reform to ensure that the market forces of affordability, promotion and availability act in a strong and directly inverse relation to the hazard of the nicotine product, and that the marketing and use of nicotine products is carefully monitored to maximise public health benefit.

▶ While it may be possible to achieve this reform and consistency by a more rational application of existing regulatory frameworks, our conclusion is that the scale of the problem and the difficulties of achieving successful reform are such that the problem will be best addressed by the creation of a nicotine regulatory authority to take control of all aspects of regulation of all nicotine products.

References

1   Doll R, Peto R, Wheatley K *et al.* Mortality in relation to smoking: 40 years' observations on male British doctors. *BMJ* 1994;309:901–11.

2   Peto R, Lopez AD. Future worldwide health effects of current smoking patterns. In: Koop CE, Pearson CE, Schwarz MR (eds), *Critical issues in global health.* San Francisco: Jossey-Bass, 2001:154–61.

3   Ezzati M, Lopez AD. Estimates of global mortality attributable to smoking in 2000. *Lancet* 2003;362:847–52.

4   World Bank. *Tobacco control at a glance,* June 2003. www1.worldbank.org/tobacco/pdf/ AAG%20Tobacco%206-03.pdf (accessed 7 August 2007).

5   World Health Organization. *WHO Framework Convention on Tobacco Control.* Geneva: WHO, 2003.

6   Levy D, Gitchell J, Chaloupka F. The effects of tobacco control policies on smoking rates: a tobacco control scorecard. Calverton, MD: PIRE Working Paper, 2003.

7   Pierce JP, Gilpin EA, Emery SL *et al.* Has the California tobacco control program reduced smoking? *JAMA* 1998;280:893–9.

8   Biener L, Harris JE, Hamilton W. Impact of the Massachusetts tobacco control programme: population based trend analysis. *Br Med J* 2000;321:351–4.

9   Department of Health. *Smoking kills.* A White Paper on tobacco. London: The Stationery Office, 1998.

10   Joossens L, Raw M. The Tobacco Control Scale: a new scale to measure country activity. *Tob Control* 2006;15:247–53.

11   Jarvis MJ. Monitoring cigarette smoking prevalence in Britain in a timely fashion. *Addiction* 2003;98:1569–74.

12   Taylor T, Lader D, Bryant A, Keyse L, McDuff TJ. *Smoking-related behaviour and attitudes, 2005.* London: Office for National Statistics, 2006.

13   Jarvis MJ, Wardle J, Waller J, Owen L. Prevalence of hardcore smoking in England, and associated attitudes and beliefs: cross sectional study. *BMJ* 2003; 326:1061–3.

14   Jarvis MJ, Wardle J. Social Patterning of individual health behaviours: the case of cigarette smoking. In: Marmot M, Wilkinson R (eds), *Social determinants of health.* Oxford: Oxford University Press, 1999.

15   Silagy C, Lancaster T, Stead L, Mant D, Fowler G. Nicotine replacement therapy for smoking cessation. *Cochrane Database Syst Rev* 2004;(3):CD000146.

16   Anderson JE, Jorenby DE, Scott WJ, Fiore MC. Treating tobacco use and dependence: an evidence-based clinical practice guideline for tobacco cessation. *Chest* 2002;121:932–41.

17   West R, McNeill A, Raw M. Smoking cessation guidelines for health professionals: an update. *Thorax* 2000;55:987–99.

18   Medioni J, Berlin I, Mallet A. Increased risk of relapse after stopping nicotine replacement therapies: a mathematical modelling approach. *Addiction* 2005;100:247–54.

19   Siahpush M, McNeill A, Hammond D, Fong GT. Socioeconomic and country variations in knowledge of health risks of tobacco smoking and toxic constituents of smoke: results from the 2002 International Tobacco Control (ITC) Four Country Survey. *Tob Control* 2006;15(Suppl 3):65–70.

20   Nelson DE, Mowery P, Tomar S *et al.* Trends in smokeless tobacco use among adults and adolescents in the United States. *Am J Public Health* 2006;96:897–905.

21   Levy DT, Mumford EA, Cummings KM, Gilpin EA, Giovino GA *et al.* The potential impact of a low-nitrosamine smokeless tobacco product on cigarette smoking in the United States: estimates of a panel of experts. *Addict Behav* 2006;31:1190–1200.

22   Lewis S, Arnott D, Godfrey C, Britton J. Public health measures to reduce smoking prevalence in the UK: how many lives could be saved? *Tob Control* 2005;14:251–4.

23   Department of Health. *Choosing health: making healthy choices easier.* London: DH, 2004.

13 Key conclusions and recommendations

Use of tobacco in society

▶ Tobacco use originated in the American continent thousands of years ago and has spread to the rest of the world in the past 500 years.

▶ Tobacco continues to be used in many different ways, but the most common form of consumption is now the cigarette.

▶ The current global epidemic of cigarette smoking is a recent phenomenon, dating from around the turn of the 20th century.

▶ The onset of the smoking epidemic typically occurs in men before women, with epidemic increases in deaths caused by smoking occurring 20–30 years after the onset of smoking.

▶ In some developed countries, both smoking prevalence and mortality rates are now falling.

▶ Most countries are at an earlier stage of the epidemic, and globally both smoking rates and mortality are rising.

▶ Smoking is the biggest avoidable cause of premature death and disability in most developed countries, and with the evolution of the global smoking epidemic will be equally important in the future wherever smoking becomes prevalent.

▶ In 2001, smoking caused 4.8 million deaths, equivalent to about one in every 12 of all deaths, globally.

▶ By 2025 there will be an estimated 1.6 billion smokers in the world, and smoking will cause approximately 10 million deaths each year.

▶ Most of these deaths will occur in people who already smoke, rather than those who start smoking between now and 2025. Therefore, whilst preventing the uptake of smoking is crucially important to the prevention of deaths in the longer term, promoting smoking cessation has a greater effect on mortality in the shorter term.

▶ It is, therefore, crucial to find ways of helping existing smokers to quit smoking, as well as preventing the uptake of smoking.

Mechanisms of action of nicotine in the brain

▸ Nicotine targets receptors whose natural function is to interact with the neurotransmitter acetylcholine.

▸ By activating these nicotinic receptors, nicotine increases the firing rate of neurons and increases the release of various neurotransmitters.

▸ The effect of nicotine on different nicotinic receptors is dose dependent and is also modified by sustained exposure, which causes some receptors to become desensitised.

▸ Long-term exposure to nicotine also causes an increase in the number of nicotinic receptors in the brain, but it is not clear how many of these receptors are functional.

▸ In animal studies, the acute reinforcing effects of nicotine appear to be dependent on dopamine release in the brain. However, nicotine-induced dopamine release is markedly curtailed or even disappears when animals are chronically exposed to the drug.

▸ Although cigarette smoking in humans promotes dopamine release, the contribution of this effect to sustained smoking behaviour is still not fully understood.

▸ It is also not clear which nicotinic receptor subtypes are responsible for the reinforcing effects of this drug in humans.

Experimental evidence on addiction to nicotine

▸ Drug dependence has commonly been perceived as being primarily a 'drug-seeking' behaviour in which an addicted individual exhibits a powerful desire or craving for the drug, and that on each occasion the addicted individual takes the drug, he or she experiences a powerful rewarding effect which is mediated directly by the drug itself.

▸ Evidence from both animal and human studies indicates that addiction to tobacco smoking is more complex. It also seems to depend critically upon the ability of nicotine to confer powerful rewarding properties on other sensory cues arising from the process of smoking, and possibly also on the circumstances in which smoking occurs.

▸ The direct reinforcing properties of nicotine are experienced predominantly only after periods of temporary abstinence, after sleep for example. For much of the remainder of the smoking day, nicotinic receptors are desensitised and the nicotine inhaled by the smoker probably does not then cause stimulation of the pathways implicated in the development of dependence. During these periods, smoking is probably reinforced primarily by conditioned stimuli (sensory cues) present in the smoke.

▶ By smoking in this way, smokers maintain their blood nicotine level at a concentration sufficient to prevent the aversive consequences of withdrawal, while continuing to derive some positive reinforcement from the conditioned stimuli present in the smoke, and, possibly, other behavioural cues associated with smoking.

▶ Hypotheses concerning the neurobiological mechanisms have been derived predominantly from studies with experimental animals. As far as it has been possible to test them, the pharmacological responses to nicotine, inhaled by humans in tobacco smoke, elicit similar effects in the human brain to those observed in experimental animals.

▶ The interplay between the pharmacological properties of nicotine inhaled in tobacco smoke and the cues and conditioned stimuli associated with human tobacco dependence are likely to be even more complex than those revealed by animal studies. Future studies which seek to establish better treatments for tobacco addiction might usefully focus on this aspect of the problem.

Nicotine addiction in humans

▶ Addiction to nicotine arises from a combination of genetic, environmental and pharmacological factors, but the characteristics of the nicotine delivery system are also crucially important.

▶ Cigarettes are the most addictive tobacco product.

▶ Cigarettes and many other tobacco products have been specifically designed, engineered and marketed to enhance both development and maintenance of addiction.

▶ Medicinal nicotine products are designed and marketed to minimise their addiction potential.

▶ The development of addiction includes changes in brain structure and function that result in cessation-associated withdrawal effects that typically persist for many weeks or longer in some individuals, thereby impairing the ability to achieve and sustain abstinence.

▸ Treatment of dependence and withdrawal can restore brain function, mood, and cognitive abilities, and thereby support cessation, but individuals appear to vary widely in how long they may require treatment, and probably in what forms of treatment are acceptable and effective.

▸ However, some of the changes in brain structure and function in smokers, particularly those who began smoking when very young, may not be entirely reversible.

▸ Hence, some smokers may never fully overcome their addiction, or even ever be able to quit all nicotine use.

Sources of nicotine

▸ There is a wide variety of nicotine products available for use, delivering a range of nicotine doses. The cigarette is the most widely used product.

▸ Cigarettes deliver high doses of nicotine into the lungs, where it is absorbed rapidly and transported directly in the systemic circulation to the brain.

▸ The nicotine from cigarettes is carried in smoke which contains thousands of other chemicals, including many that are carcinogenic or otherwise toxic.

▸ Some of these toxins are present in tobacco before combustion. Most are combustion products.

▸ Smokeless tobacco also contains toxins and carcinogens, but delivers high doses of nicotine without most of the toxic components in smoke.

▸ Medicinal nicotine products deliver pure nicotine, but in relatively low doses and, particularly for nicotine transdermal patches, very slowly. They do not deliver other toxic chemicals.

▸ The available alternative nicotine products all deliver nicotine more slowly than cigarettes, and are, therefore, probably less addictive.

▸ It is possible that alternative nicotine products could provide a safer long-term substitute for cigarette smoking. If so, this could benefit individual and public health.

The risks of smoking

▸ Smoking currently kills nearly 5 million people each year.

▸ In the 20th century, there were an estimated 100 million premature deaths attributable to smoking. If current smoking patterns continue, there could be more than 1 billion deaths in the 21st century.

▸ A large proportion of the population in many countries still take up smoking when young and continue their habit into middle and old age. Of all those who die from smoking worldwide, half are in developing countries, but this proportion is likely to increase unless the smoking epidemic can be halted.

▸ Passive smoking is also a major avoidable cause of death and disability.

▸ Stopping smoking is highly effective, even in older smokers, and generates immediate benefits to health.

▸ To avoid a greater public health disaster in the current century, more efforts should be made to prevent non-smokers from starting to smoke and to encourage smokers to quit.

The risks of medicinal nicotine

▸ Extensive experience with nicotine replacement therapy (NRT) in clinical trial and observational study settings demonstrates that medicinal nicotine is a very safe drug.

▸ Adverse effects are primarily local and specific to the mode of delivery used.

▸ NRT does not appear to provoke acute cardiovascular events, even in people with pre-existing cardiovascular disease.

▸ There is no direct evidence that NRT therapy is carcinogenic, or influences the risk of other common smoking-related diseases in humans.

▸ Evidence on the safety of NRT during pregnancy is limited, but suggests that NRT does not increase the risk of major developmental anomalies or reduce birth weight. However, NRT may increase the risk of minor musculoskeletal anomalies. Further evidence on these effects is needed.

▸ Evidence on the safety of long-term use of NRT is lacking, but there are no grounds to suspect appreciable long-term adverse effects on health.

▶ In any circumstance, the use of NRT is many orders of magnitude safer than smoking.

The risks of smokeless tobacco

▶ Smokeless tobacco is not a single product, but rather a summary term for a range of different tobacco products which deliver nicotine without combustion.

▶ Smokeless tobacco products differ substantially in their risk profile in approximate relation to the content of toxins in the tobacco.

▶ In some parts of the world (particularly South Asia), smokeless tobacco is commonly mixed with other products that are themselves harmful.

▶ On toxicological and epidemiological grounds, some of the Swedish smokeless (snus) products appear to be associated with the lowest potential for harm to health.

▶ Swedish smokeless products appear to increase the risk of pancreatic cancer, and possibly cardiovascular disease, particularly myocardial infarction.

▶ Some smokeless tobacco products also increase the risk of oral cancer, but, if true of Swedish smokeless tobacco, the magnitude of this effect is small.

▶ All of the above hazards of smokeless tobacco are of a lower magnitude than those associated with cigarette smoking.

▶ Smokeless tobacco products have little or no effect on the risk of chronic obstructive pulmonary disease or lung cancer.

▶ Therefore, in relation to cigarette smoking, the hazard profile of the lower risk smokeless products is very favourable.

▶ Smokeless tobacco use by pregnant women is harmful to the unborn fetus, but the hazard of smokeless use relative to maternal cigarette smoking is not clearly established.

▶ In Sweden, the available low-harm smokeless products have been shown to be an acceptable substitute for cigarettes to many smokers, while 'gateway' progression from smokeless to smoking is relatively uncommon.

▸ Smokeless tobacco therefore has potential application as a lower hazard alternative to cigarette smoking.

▸ The applicability of smokeless tobacco as a substitute for cigarette smoking if made available to populations with no tradition of smokeless use is not known.

Current approaches to nicotine product regulation

▸ Nicotine product regulation has developed in a largely reactive and piecemeal fashion over the years.

▸ Smoked tobacco products remained free from regulation for many years, and are now subject to minimal controls on content, delivery and safety.

▸ Some smokeless tobacco products are regulated strictly (they are prohibited) whilst others are subject to even less regulation than cigarettes.

▸ Medicinal nicotine products are regulated strictly, as medicines.

▸ The lax regulation of most tobacco products affords considerable market freedom for tobacco companies to innovate and develop their products.

▸ The tight regulation of medicinal nicotine imposes very strict restrictions on new product development.

▸ Some newly launched tobacco products, including the potential reduced exposure products (PREPs), seem to lie completely outside the current regulations.

▸ This clear and unjustifiable regulatory imbalance works against public health.

▸ UK government resources dedicated to tobacco product regulation are very small.

▸ History demonstrates that regulatory change can achieve substantial changes in consumption of different tobacco products.

▸ The regulation of nicotine products needs to be radically overhauled to encourage the use of less harmful products and reduce the use of the more harmful sources of nicotine.

▸ Whilst some progress can be made in this regard through the existing regulatory systems, the establishment of a nicotine and tobacco regulatory authority is the preferred way of bringing comprehensive and rational controls on the nicotine product market that will minimise the harm caused by nicotine use.

Current nicotine product use and socioeconomic deprivation

▸ In Western industrial society, and particularly in the UK, cigarette smoking is strongly linked with socioeconomic disadvantage.

▸ People in disadvantaged socioeconomic groups are more likely to smoke, to smoke heavily and to be more heavily addicted to smoking.

▸ This is also true of other disadvantaged groups, such as people with mental health problems and people in prison.

▸ Smokers in disadvantaged socioeconomic groups are just as likely to want to quit smoking, and more likely to use smoking cessation services, than relatively advantaged smokers.

▸ Smokers in disadvantaged socioeconomic groups are much less likely to succeed in quitting smoking.

▸ As a result, and in contrast to more advantaged groups, smoking prevalence has changed very little in recent years among the most disadvantaged in society.

▸ Smoking therefore causes even more death and disability in disadvantaged groups than in the rest of society, and is indeed the biggest cause of social inequalities in health.

▸ Children growing up in disadvantaged households are more likely to be exposed to cigarette smoke in the home, and are more likely to start smoking and to start when especially young.

▸ Disadvantaged smokers, and their children, have the most to gain from harm reduction strategies for tobacco use.

Ethics, human rights and harm reduction

▸ Although stopping tobacco use is the ideal outcome for individual and public health, this is often difficult to achieve. Making a wider range of safer products available would be a harm reduction approach to tobacco control.

▸ Harm reduction approaches in public health are sometimes criticised for condoning the activity they are trying to make safer.

The Royal College of Physicians takes no position on the morality of smoking. However, since smoking is dangerous to health and is hard to give up, the College wants to see a range of effective methods to help smokers quit or to reduce the harm they sustain.

▶ The present status quo, in which cigarettes are freely available, medicinal nicotine products are available but under regulations that restrict availability and effectiveness, and some smokeless tobacco products are prohibited, denies smokers the right to choose safer nicotine products.

▶ Balancing the nicotine market, so that all nicotine products are equally available and comparably priced, would provide smokers with choice but would not encourage change from high-risk to lower-risk products.

▶ Rebalancing the market in favour of the safest nicotine products would provide choice, encourage safer nicotine use, and reduce morbidity and mortality.

▶ The ethical aspects of regulating alternatives to smoking tobacco are complex, and three positions can be defended: maintaining the status quo; making alternatives to smoking tobacco as easily available as smoking tobacco now is; and making alternatives more easily available than smoking tobacco now is.

▶ Each approach represents a balance of consumer rights, consumer protection, fairness and general policy considerations.

▶ The College favours an approach which would make smoking easier to give up, while discouraging starting smoking in the first place, through making alternatives to smoking tobacco more widely available and through regulating smoking products more tightly by pricing, marketing controls, and formal regulation of the production and sale of tobacco products.

Implications for health policy and nicotine product regulation

▶ Most of the deaths and disease caused by smoking in the near- and medium-term future will occur in people who are smoking now.

▶ Current conventional preventive measures focus entirely on preventing uptake of smoking and helping smokers to quit smoking.

▸ This approach will be ineffective for the millions of smokers who, despite best efforts to persuade and help them to quit, will carry on smoking. Half of these smokers, representing millions in the UK alone, will die as a result.

▸ This burden of mortality and morbidity will markedly exacerbate social inequality in health.

▸ Tobacco control policy needs to be radically extended to address the needs of these smokers with implementation of effective harm reduction strategies.

▸ Harm reduction in smoking can be achieved by providing smokers with safer sources of nicotine that are acceptable and effective cigarette substitutes.

▸ There is a moral and ethical duty to provide these products to addicted smokers.

▸ Current systems of regulation of nicotine products inhibit the development of innovative medicinal nicotine substitutes for cigarettes and perpetuate the use of the most dangerous nicotine products. This is unjust, irrational and immoral.

▸ Nicotine product regulation must therefore be reformed.

▸ The unprecedented and unjustifiable market freedoms enjoyed by manufacturers of cigarettes and other smoked tobacco products must end.

▸ The development of new, more effective, more acceptable and user-friendly medicinal nicotine substitutes for smoking needs to be encouraged.

▸ A major step towards an effective harm reduction strategy could be taken by the implementation of simple changes to the regulation, promotion and taxation of medicinal nicotine.

▸ Low nitrosamine smokeless tobacco products may have a positive role to play in a coordinated and regulated harm reduction strategy which maximises public health benefit and protects against commercial market exploitation.

▸ The regulation of nicotine products, whether medicinal or tobacco-based, thus needs radical reform to ensure that the market forces of affordability, promotion and availability act in a strong and directly inverse relation to the hazard of the nicotine product, and that the marketing and use of nicotine products are carefully monitored to maximise public health benefit.

▶ While it may be possible to achieve this reform and consistency by more rational application of existing regulatory frameworks, our conclusion is that the scale of the problem, and the difficulties of achieving successful reform, are such that the problem will be best addressed by the creation of a nicotine regulatory authority to take control of all aspects of regulation of all nicotine products.